MEMOIRS

of the
American Mathematical Society

Number 451

Effective Algebraic Topology

Rolf Schön

July 1991 • Volume 92 • Number 451 (end of volume) • ISSN 0065-9266

American Mathematical Society
Providence, Rhode Island

1980 *Mathematics Subject Classification* (1985 *Revision*).
Primary 18E10, 55T10, 55Q05, 55P35, 55S45;
Secondary 18G40, 18-04, 55-04, 55R20, 55R05.

Library of Congress Cataloging-in-Publication Data

Schön, Rolf, 1943–
    Effective algebraic topology/Rolf Schön
    p. cm. – (Memoirs of the American Mathematical Society; no. 451)
    Includes bibliographical references.
    ISBN 0-8218-2522-4
    1. Algebraic topology. I. Title. II. Series.
QA612.S36 1991   91-15008
514′.2–dc20   CIP

Subscriptions and orders for publications of the American Mathematical Society should be addressed to American Mathematical Society, Box 1571, Annex Station, Providence, RI 02901-1571. *All orders must be accompanied by payment.* Other correspondence should be addressed to Box 6248, Providence, RI 02940-6248.

SUBSCRIPTION INFORMATION. The 1991 subscription begins with Number 438 and consists of six mailings, each containing one or more numbers. Subscription prices for 1991 are $270 list, $216 institutional member. A late charge of 10% of the subscription price will be imposed on orders received from nonmembers after January 1 of the subscription year. Subscribers outside the United States and India must pay a postage surcharge of $25; subscribers in India must pay a postage surcharge of $43. Expedited delivery to destinations in North America $30; elsewhere $82. Each number may be ordered separately; *please specify number* when ordering an individual number. For prices and titles of recently released numbers, see the New Publications sections of the NOTICES of the American Mathematical Society.

BACK NUMBER INFORMATION. For back issues see the AMS Catalogue of Publications.

MEMOIRS of the American Mathematical Society (ISSN 0065-9266) is published bimonthly (each volume consisting usually of more than one number) by the American Mathematical Society at 201 Charles Street, Providence, Rhode Island 02904-2213. Second Class postage paid at Providence, Rhode Island 02940-6248. Postmaster: Send address changes to Memoirs of the American Mathematical Society, American Mathematical Society, Box 6248, Providence, RI 02940-6248.

Contents

Abstract

Four papers introduce and develop a new method for effective computations in Algebraic Topology. It applies to such sectional areas, where groups of interest appear as terms of exact sequences, which are direct limits of computable, exact or non-exact sequences. The general case of application is the computation of the value $F(X)$ of a functor F on the direct limit X of an algorithmically given sequence of (dimensionwise) finite complexes $_iX$, $i \geq 1$, with $F(_iX)$ computable. For example a Kan-envelope X of a given, finitely generated complex $\underline{X} = {}_1X$ can be constructed in this way, and also many other complexes derived from X, e.g. loop complexes LX or Postnikov-stages $n.X$. Spectral sequences provide a host of instances of functors F, and make the homotopy theory of finitely generated, simply connected complexes largely accessible to effective computations. Their homotopy groups can be computed via the homology of iterated loop complexes. Their Postnikov-stages have computable (co-)homology groups and k-invariants.

Preface

This work gives the first four of a series of papers, which provide means for effective solutions to computation, construction, and decision problems in Algebraic Topology. Restrictions of storage and time are ignored, but one section (the last one of paper 3) suggests further developments towards more practicability.

The novelty lies on a rather elementary level. It consists in a new handling of exact sequences. Appropriately restricted by algorithmical conditions they allow one to compute terms of interest effectively from other known terms. The wide spread occurrence of exact sequences and the naturalness of these conditions opens a wide range of application.

Here we consider exact sequences, which are direct limits of algorithmically given exact or non-exact sequences. The **extended calculability** of individual terms is the basic concept introduced in paper 1 and studied with respect to its Serre-class property. Spectral sequences of **fibrations**, which are approximated by maps between finite complexes, yield diagrams of such exact sequences. The main result of paper 2 is that each homology group of one of the complexes $F \to E \to B$ is extended calculable if each one of the other two is extended calculable. In particular, we can consider **path-fibrations** $PX \to X$ and **Postnikov-stages** $n.X \to n-1.X$, which are associated to an effectively constructed Kan-envelope $X = \varinjlim(_iX)$, $_1X \subset {}_2X \subset \ldots \subset {}_iX \subset \ldots$, of a finitely generated complex $_1X$. Paper 3 makes use of the first case. Its main result is the effective computability of the **homotopy groups** of finite, simply connected complexes achieved via the homology of iterated loop spaces. Paper 4 considers the Postnikov-stages of such complexes and proves the computability of their (co-)homology groups and **k-invariants**.

The papers are revisions of first versions written up since the end of 1983. Most material had been brought to the knowledge of interested topologists by many mailed manuscripts and several talks abroad. I thank for kind invitations and hospitality.

Rolf Schön
Heidelberg, May 1990

A Five Lemma for Calculations in Homological Algebra

Abstract. As a rule, exact sequences in homological algebra do not give complete information about an object one wants to calculate. This note shows, that at least in principle full information can be received, if homological algebra is restricted to finitely generated abelian groups, and the notation of calculability is substantially strengthened. Roughly speaking, we get then the result, that the middle term of a 5-term exact sequence is calculable, if the other four are. As a first step towards applications we prove a calculability proposition for spectral sequences or maps. More generally, an application of the offered Five-Lemma Algorithm can be taken into account, when one wants to calculate the colimit of an algorithmically given (but possibly at no time completely known) sequence of finitely generated abelian groups, and it is known (by a diagram of interlocking exact sequences), that this colimit is finitely generated.

A fundamental lemma for algorithmic calculations in homological algebra will be proved; see the 5. Lemma below. The final section gives an application to spectral sequences of maps. The treatment is fitting to the elementary nature of the topic.

I The Five-Lemma Algorithm

AB denotes the category of sequences (L) of homomorphisms $L(i) \to L(i+1), i \geq 1$, between finitely generated abelian groups $L(i)$, $i \geq 1$, with finitely generated colimit $L = \lim_{\longrightarrow} L(i)$. A morphism $(L) \to (M)$ is a sequence of homomorphisms $L(i) \to M(i)$, $i \geq 1$, such that the squares

$$
\begin{array}{ccc}
L(i) & \to & L(i+1) \\
\downarrow & & \downarrow \\
M(i) & \to & M(i+1)
\end{array}
$$

are commutative. The following lemma motivates the notion of calculability we need, and is at the same time the basic fact, which forces the termination of our algorithms.

1. Lemma   (Notation: $L(i,j) = \text{image}(L(i) \to L(j))$, if $i \leq j$).
For each object (L) in AB there exist integers $j \geq i \geq 1$, such that (the canonical map) $L(j) \to L$ induces an isomorphism $L(i,j) \to L$.

Proof
By exactness of the colimit (of sequences), the images of $L(k) \to L, k \geq 1$ form an ascending chain of subgroups of L with union L. Because L is finitely generated this chain becomes stationary, such that we have an epimorphism $L(i) \to L$ for a certain integer $i \geq 1$. Again by exactness of the colimit, the kernels of $L(i) \to L(k)$, $k \geq i$,

---

1980 Mathematics Subject Classification (1985 Revision).
Primary 18 E 10, Secondary 18 G 40, 18-04, 55-04

form an ascending chain of subgroups of $L(i)$ with union the kernel of $L(i) \to L$, which is finitely generated. Hence this chain becomes stationary, such that we have an identity kernel$(L(i) \to L(j)) =$ kernel$(L(i) \to L)$ for a certain integer $j \geq i$. Therefore $L(i,j) =$ image$(L(i) \to L(j)) \cong L(i)/$kernel$(L(i) \to L(j)) = L(i)/$kernel$(L(i) \to L) \cong$ image$(L(i) \to L) = L$.

Introductory Remarks

By the fundamental structure theorem for finitely generated abelian groups, e.g. [6, p.101], every finitely generated abelian group G can be characterized (up to an isomorphism) by a tuple

$$TG = (r; t_1, t_2, \ldots, t_n)$$

with $r \geq 0$ an integer, the free rank of G, and each t_i an integer tuple $(p_i; 1_i, 2_i, \ldots, k_i)$, where the p_i's, $1 \leq i \leq n$, are different prime numbers, and the j_i's are integers ≥ 1, such that G is isomorphic to a direct sum TG, same notation, having as summands r copies $\mathbb{Z}$, and one copy $\mathbb{Z}/(p_i^{j_i})\mathbb{Z}$ (= cyclic group of order $p_i^{j_i}$, called torsion coefficient or invariant) for each pair (p_i, j_i). Calculating G means calculating TG, the type of G. For instance G may be $H_i(K)$, the i-th homology group of a finite simplicial complex K; then from the data i and K, TG can be calculated (by known computer conform algorithms, [1, p.330] or [5]).

More generally, we need for finite complexes (= simplicial sets) $K \subset L$ the computability of the homology groups $H_i(L, K)$ and of the connecting homomorphisms $H_i(L, K) \to H_{i-1}(K)$. These facts are taken for granted.

Our presentation of algorithms is often informal in so far, as declared INPUT-data have to be completed by obviously available data given beforehand.

Definitions
An object (L) in AB is called **precalculable**, if there exists an algorithm, e.g. [13, p.131], which assigns to an integer $i \geq 1$ the types $TL(i)$, $TL(i+1)$, and a homomorphism $TL(i) \to TL(i+1)$ (numerically representable by an integer matrix), such that the object (TL) is isomorphic to (L) in AB. Comment: In case the sequence (L) is finite, i.e. if $L(i) \xrightarrow{=} L(i+1)$ for all i bigger than some $k \geq 0$, then "precalculable" coincides with "effectively calculable" (= "calculable" in this note). (L) is called epi- (mono-, iso-) calculable, briefly **epical (monocal, isocal)** if there exists an algorithm, which assigns to an integer $i \geq 1$ a pair of integers j, k with $i \leq j \leq k$, such that one of the following equivalent conditions (1), (2) is fulfilled.

(1) $L(k) \to L$ induces an epi- (mono-, iso-) morphism $L(j, k) \to L$.
(2) For each $l \geq k$ $L(l) \to L$ induces an epi- (mono-, iso-) morphism $L(j, l) \to L$.

Comment: The equivalence (1) $\Leftrightarrow$ (2) is immediate because of the factorization $L(j, k) \to L(j, l) \to L$, where the first map is an epimorphism. $j = {'a(i)}$ resp. $k = a'(i)$ is called the **early resp. late epimorph** (monomorph, isomorph) index associated to i, and the assignment $i \to ({'a(i)}, a'(i))$ is then (by usual definition) the associated algorithmic function.

Obvious facts: (3) If (L) is epical, then it is epical with the same early and late epimorph index. (4) If (L) is isocal, then it is epical and monocal. (5) If (L) is isocal, then with k a late isomorph index $L(k) \to L$ is a split epimorphism. k is then called **split epimorph index.** (L) is called **extended calculable**, if the colimit L of (L) and an epimorph integer i (= an integer i, such that $L(i) \to L$ is epimorphic) is calculable. A pair (i, TL) with i an epimorph integer, and L, the colimit of (L), is called an **extended value** of (L).

2. Lemma

An object (L) in AB is epical, if and only if an epimorph integer is calculable.

Proof

Let $'j$ be a calculated epimorph integer, then the assignment $i \to (j, j)$ with $j = \max\{'j, i\}$ is an algorithmic function for an epicalculation because of the factorization $L('j) \to L(j) \to L$. The converse is trivial: choose a calculated epimorph index as epimorph integer.

3. Lemma

For a precalculable object (L) in AB, the following properties are equivalent:

(a) (L) is isocal
(b) (L) is epical and monocal
(c) (L) is extended calculable
(d) An extended value of (L) is calculable.

Proof

The implication (a) $\Rightarrow$ (b) is the obvious fact (4) above. (b) $\Rightarrow$ (a): Let $i \to ('e(i), e'(i))$, $i \to ('m(i), m'(i))$ be the algorithmic functions for an epi- resp. monocalculation. The assignment $i \to ('m(j), m'(j))$ with $j = 'e(i)$ is an algorithmic function for an isocalculation because of the factorization $L('e(i)) \xrightarrow{u} L('m(j), m'(j)) \xrightarrow{v} L$, for vu epi and v mono implies v iso.

    (a) $\Rightarrow$ (c): Let $i \to ('a(i), a'(i))$ be the algorithmic function for an isocalculation. We have the calculable epimorph integer $'a(1)$ and an isomorphism $L('a(1), a'(1)) \xrightarrow{\cong} L$. Because L is precalculable, $L('a(1), a'(1))$, as image of a calculable homomorphism between calculable finitely generated abelian groups, is calculable. Hence L is extended calculable.

    (c) $\Rightarrow$ (a): Let $'j$ be a calculated epimorph integer, and $i \geq 1$ a given integer. Then $j = \max\{'j, i\} \geq i$ is also an epimorph integer because of the factorization $L('j) \to L(j) \to L$. We choose $j = 'a(i)$ as early isomorph index, and calculate a late isomorph index:

INSTRUCTION: Run through the integers $k \geq 'a(i)$, and calculate $L('a(i), k)$.
                  Stop, if $L('a(i), k)$ has the type TL.

By the second part of the proof of the 1. Lemma such an integer k exists, and hence the calculation process stops after finitely many steps. In case it stops, say at $k = a'(i)$, the canonical map $L('a(i), a'(i)) \to L$ is an epimorphism between isomorphic, finitely generated abelian groups, and hence an isomorphism by an elementary property of such groups, e.g. [6, p.42]. Therefore $i \to ('a(i), a'(i))$ is an algorithmic function for an isocalculation.

The equivalence (c) $\Leftrightarrow$ (d) is almost identical with the definition of "extended calculable", and links the introduced calculability notion with the classical one.

4. Lemma   (A generalization of the 1. Lemma)

Let (L_r), $1 \le r \le n$, be finitely many objects in AB, and $\underline{i}$, $i \ge 1$, a sequence of n-tuples $\underline{i} = (i_1, i_2, \ldots, i_n)$ of integers $i_j \ge 1$, such that the sequence of the minima $\min\{i_1, i_2, \ldots, i_n\}$ is unbounded (i.e. for each integer $k \ge 1$ there exists an n-tuple $(i_1, i_2, \ldots i_n)$, such that $k \le i_j$ for each j, $1 \le j \le n$). Then among the set of $(n+1)$-tuples

$$T = \{(\underline{i}, l) / l \ge a_i\}\,, \quad \text{with } a_i \ge \max\{i_1, i_2, \ldots, i_n\} \quad \text{any integer sequence,}$$

there exists some one, say $(\underline{j}, k)$, such that the canonical maps $L_r(j_r, k) \to L$ are isomorphisms for all r, $1 \le r \le n$.

Proof   (See proof of the 1. Lemma)

By the unboundedness condition, there exists an n-tuple j, such that $L_r(j_r) \to L$, $1 \le r \le n$, are epimorphisms. As in the proof of the 1. Lemma, there exist integers $k_r \ge j_r$, such that $L(j_r, k_r) \to L$ are isomorphisms, $1 \le r \le n$. Choose $k = \max\{k_1, k_2, \ldots, k_n, a_j\}$, then $(\underline{j}, k) \in T$, and the canonical maps $L_r(j_r, k) \to L$, $1 \le r \le n$, are isomorphisms (with regard of the implication (1) $\Rightarrow$ (2) within the isocal definition above).

Definitions

Let us denote by $AB\text{-}n$ the category of n-term sequences $"L"$ of objects in AB, i.e.

$$"L" = ((L_n) \to (L_{n-1}) \to \ldots \to (L_1))\,, \quad (L_r) \in AB\,, \quad 1 \le r \le n\,.$$

$"L"$ is called a **chain complex**, if for each $i \ge 1$, $L_n(i) \to \ldots L_1(i)$, is a chain complex in the common sense.

$"L"$ is called **precalculable**, if there exists an algorithm, which assigns to integers $i \ge 1$ n-term sequences of types of finitely generated, abelian groups

$$TL(i) = (TL_n(i) \to TL_{n-1}(i) \to \ldots \to TL_1(i))\,,$$

and homomorphisms $TL(i) \to TL(i+1)$, such that the object $"TL"$ is isomorphic to $"L"$ in $AB\text{-}n$.

5. Lemma   (the Five-Lemma Algorithm)

Let

$$"L" = (("L) \to ('L) \to (L) \to (L') \to (L''))$$

be a precalculable, 5-term chain complex in AB, such the colimit chain complex $"L"$, of $("L")$, is exact, then the following is true:

(a) If $('L), (L')$ epical, (L'') monocal, then (L) is epical.

(b) If $('L), (L')$ monocal, $(''L)$ epical, then (L) is monocal.

(c) If $('L), (L')$ isocal, (L'') monocal, $(''L)$ epical, then (L) is isocal.

Proof

(a) Let $j \to ('a(j),' b(j))$, $j \to (a'(j), b'(j))$, $j \to (a''(j), b''(j))$ be the algorithmic functions, which are associated to $'L, L'$, and L'' respectively by assumption. We have then the composed algorithmic function

$$j \to b(j) = \max\{'b(j), b'('a(j)), b''a'('a(j))\} \ .$$

Let $i \geq 1$ be a given integer.

INSTRUCTION: Run through the set of index pairs $M_i = \{(j, k)/k \geq b(j), j \geq i\}$
and decide, whether the chain complex

$$C_{j,k} : {}'L('a(j)k) \to L('a(j), k) \to L'(a'(a'(j)), k) \to L''(a''a'('a(j)), k)$$

is exact. Stop, if this is the case.

Because $'a(j) \leq a'('a(j)) \leq a''a'('a(j))$ $C_{j,k}$ is well defined, and, because $''L''$ is precalculable, $C_{j,k}$ is a calculable chain complex in the category of finitely generated abelian groups (where exactness is decidable). By the 4. Lemma a 5-tuple $('a(j),'a(j), a'('a(j)), a''a'('a(j)), k)$ with $(j, k) \in M_i$, and $C_{j,k}$ exact, exists (here clearly is used, that a termwise isomorphism $C_{j,k} \to {}''L''$ onto the exact colimit sequence $''L''$ implies the exactness of $C_{j,k}$); therefore the calculation process stops after finitely many steps. In case it stops, say at the pair (j, k), we have a chain homomorphism $C_{j,k} \to {}''L''$ between exact sequences, and can apply the classical Five Lemma, e.g. [4, p.14], which yields an epimorphism $L('a(j), k) \to L$ (regard, that $k \geq b(j)$, and we have the implication (1) $\Rightarrow$ (2) within the epical (monocal) definition above). Therefore the assignment $i \to (a(i), b(i)) = ('a(j), k)$ is an algorithmic function for an epicalculation of L.

(b) The proof of (a) is written down, such that we have a word for word analogy. We just give the corresponding instruction.

INSTRUCTION: Run through the set of index pairs $\bar{M}_i = \{(j, k)/k \geq \bar{b}(j), j \geq i\}$
with $\bar{b}(j) = \max\{''b(j),' b''a(j), b'('a''a(j))\}$, and decide whether
the chain complex

$$D_{j,k} : {}''L(''a(j), k) \to {}'L('a''a(i), k) \to L('a''a(i)k) \to L'(a'('a''a(i))k)$$

is exact. Stop, if this is the case.

(c) is obvious because of (a) and (b) and the equivalence (a) $\Leftrightarrow$ (b) of the 3. Lemma.

Remarks

1. The proof of the 5. Lemma yields a somewhat involved, composed algorithm. But in case of an epicalculation, the proof of the 2. Lemma gives a translation into an equivalent uncomposed algorithm, which (uses a single calculated epimorph integer and) is much simpler. The same is true in case of an isocalculation by using the implications (a) $\Rightarrow$ (d) $\Rightarrow$ (a) of the 3. Lemma. These translations should be

made, when the whole algorithm (and not only single calculation result) is needed for further calculations. (A similar simplification for monocalculations is probably not possible without restricting the class of groups we are handling.)

2. Not only for complexity reasons, but also for an application in [2] it is useful to calculate minimal indexes before entering further calculations. This can be done in the case of an isocalculation as follows. Let j be an early and k a late isomorph index for L. A minimal early index is calculated by the following instruction.

INSTRUCTION: Run through the (finite) decreasing sequence of integers $i, 1 \leq i < j$, and decide whether $L(i) \to L(j, k)$ is epimorphic. Stop, if this is not the case.

If the calculation process stops at an integer $i > 1$, we choose $m = i + 1$ as minimal early isomorph index, and else, i.e. with $i = 1$, we choose $m = 1$ or $m = 2$ according to $L(1) \to L(j, k)$ is epimorphic or not.

A minimal late isomorphic index can then be calculated by the following instruction.

INSTRUCTION: Run through the (finite) increasing sequence of integers $'j, m \leq 'j \leq k$, and decide, whether $L(m, 'j) \to L(j, k)$ is an isomorphism. Stop, if this is the case.

If the calculation process stops at $'j = n$, we choose n as minimal late isomorph index.

II Spectral Sequences of Filtrations, see [4, p.332]

Let X be a simplicial set (= complex, e.g. [3, p.5]), and $i/X \subset i+1/X$, $i \subset \mathbb{Z}$, a filtration of subcomplexes $i/X \subset X$ with $i/X = \emptyset$ for $i < 0$, and with union X. We have the associated spectral sequence, which is by definition the sequence of bigraded abelian groups

$$/n, p, r/ =$$
$$_nE_p^r(X) = \text{image}(H_n(p/X, p - r/X) \to H_n(p + r - 1/X, p - 1/X)), \, r \geq 1,$$

together with the differentials

$$d^r = {_nd_p^r} : /n, p, r/ \to /n - 1, p - r, r/, \quad r \geq 1,$$

which are the restrictions of the connecting homomorphisms

$$H_n(p + r - 1/X, p - 1/X) \to H_{n-1}(p - 1/X, p - r - 1/X)$$

onto $/n, p, r/$. The inclusion $(p + r - 1/X, p - 1/X) \to (p + r/X, p - 1/X)$ induces the homomorphism i in the diagram

$$
\begin{array}{c}
(p + r, p) \overset{d}{\to} (p, p - 1) \to (p + r, p - 1) \\
d \searrow \quad \nearrow \quad\quad \searrow \quad i \nearrow \\
(p, p - r) \overset{j}{\to} (p + r - 1, p - 1) \\
k \nearrow \quad \searrow \quad\quad \nearrow \quad \searrow d \\
(p, p - r - 1) \to (p, p - 1) \underset{d}{\to} (p - 1, p - r - 1)
\end{array}
$$

where the d's are connecting homomorphisms, and (p, i) stands for $H_\alpha(p/X, i/X)$ with $\alpha = n + 1$ on the left above, $\alpha = n - 1$ on the right below, and else $\alpha = n$.

It is image$(j) = /n, p, r/$, and image$(ijk) = /n, p, r + 1/$; the lower and the upper sequence is exact. By diagram chase, one shows easily, that i induces a homomorphism

$$ i' : \text{kernel}(d^r) = \text{kernel}(d) \cap /n, p, r/ \to /n, p, r + 1/ $$

by restriction, that i' is epimorphic, and that kernel(i') is contained in image$(d^r) = $ image(jd) and vice versa. Hence i induces an isomorphism

$$ \text{homology}\left(\xrightarrow{d^r} /n, p, r/ \xrightarrow{d^r}\right) \xrightarrow{\cong} /n, p, r + 1/ $$

which is a fundamental property of spectral sequences (and part of their definition, when they are introduced more generally [4, p.319]). We have indicated this simple proof, because most recent literature, see e.g. [12, §2.2.2], is still ignorant of it.

The infinity terms of the spectral sequence $/n, p, r/$ are defined by

$$ /n, p/ = {}_n E_p(X) = \text{image}(H_n(p/X) \to H_n(X, p - 1/X)) , $$

and can formally be received out of $/n, p, r/$ by putting $r = \infty$, and defining $p - \infty/X = \emptyset$. $p + \infty/X = X$ for all p. An identical diagram chase as above, now applied to the following diagram

$$
\begin{array}{ccccc}
H_n(p - 1/X) & \to & H_n(X) & \to & H_n(X, p - 1/X) \\
& \searrow\ 1\ \nearrow & & \searrow 1\ \nearrow & \\
& H_n(X) & \xrightarrow{1} & H_n(X) & \\
& \nearrow\ \searrow 1 & 1\ \nearrow & \searrow & \\
H_n(p/X) & \longrightarrow & H_n(X) & \to & H_n(X, p/X) ,
\end{array}
$$

where the 1's are identities, yields an isomorphism

$$ \text{homology}(H_n(p - 1/X) \to H_n(X) \to H_n(X, p/X)) \xrightarrow{\cong} /n, p/ $$

or equivalently the short exact sequence

$$ (0) \quad 0 \to p - 1/H_n(X) \to p/H_n(X) \to /n, p/ \to 0 $$

with $p/H_n(X) = \text{image}(H_n(p/X) \to H_n(X)) = \text{kernel}(H_n(X) \to H_n(X, p/X))$, the first identity by definition of $p/H_n(X)$, the latter one by the long exact sequence of the pair $(X, p/X)$. Hence the infinity terms $/n, p/$ are quotients of successive terms of a filtration of $H_n(X)$.

III The Spectral Sequence of a Map

(a) Let $f : X \to Y$ be a simplicial map between complexes. A filtration p/X of X is defined by

$$ p/X = f^{-1}(Y^p) = \text{preimage of the } p\text{-skeleton } Y^p \text{ of } Y , $$

and the associated spectral sequence $/n, p, r/$ is defined by **II**.

We make the following assumptions:

(b) $f : X \to Y$ is the colimit of a preconstructible sequence of maps $_if : {}_iX \to {}_iY$ between (dimensionwise) finite complexes.

(Explanation: Exactly speaking, we have a sequence of map-pairs $({}_iu, {}_iv : {}_if \to {}_{i+1}f, i \geq 1$, (i.e. of commutative squares) with colimit f in the category of simplicial maps, and furthermore an algorithm, which assigns to integers $i \geq 1$, $n \geq 0$ the square $({}_iu, {}_iv)$ restricted onto dimension n.)

(c) For a fixed given integer $r > 1$ each term $/n, p, r/$ is finitely generated.

Comments:

If a group L is the direct limit of a sequence (L), which is obvious from the context, we say in the sequel also that L is extended calculable. The following implication is a part of the implication (c) $\Rightarrow$ (d) of the 3. Lemma.

L is **extended calculable** $\Rightarrow$ L is **effectively calculable**.

One should realize this implication, when one wants to appreciate the usability of our results for concrete applications.

Proposition   (for calculating the total complex, Notation: $/n, p, r/ = {}_nE_p^r(X)$)
With (a), (b), (c) above and the following assumption (d).

(d) $/n, p, r/$ is extended calculable for each n and p,

these statements are true:

(e) each term $/n, p, r + 1/$ is extended calculable
(f) each infinity term $/n, p/$ is extended calculable
(g) each homology group $H_n(X)$ is extended calculable

Proof

Consider the exact sequences

(1) $0 \to I/n, p, r/ \to /n, p, r/$ (inclusion of an image)
(2) $/n + 1, p + r, r/ \to I/n, p, r/ \to 0$ (epimorphism d^r onto its image)
(3) $0 \to K/n, p, r/ \to /n, p, r/ \to I/n - 1, p - r, r/ \to 0$ (elementary fact, [4, p.12])
(4) $0 \to I/n, p, r/ \to K/n, p, r/ \to /n, p, r + 1/ \to 0$ (fundamental property of spectral sequences, see **II**)
where $I/n, p, r/$, and $K/n, p, r/$ denote the image resp. kernel of the differential d^r going into resp. from $/n, p, r/$, and furthermore
(5) $0 \to /n, p, r/ \to /n, p/ \to 0$ for $r \geq \max\{p + 1, n - p + 2\}$ (see (h) below).
(6) $0 \to p - 1/H_n(X) \to p/H_n(X) \to /n, p/ \to 0$ (see **II** above).

These sequences are colimits of the corresponding ones, which are associated to the spectral sequences $_i/n, p, r/$ of the maps $_if : {}_iX \to {}_iY, i \geq 1$. The preconstructibility condition (b) implies the precalculability condition of the Five-Lemma Algorithm in each case (see the introductory remark, and regard the obviously algorithmic constructibility of $p/{}_iX$ by the map $_if : {}_iX \to {}_iY$ because of the finiteness condition of (b) above).

We now permanently use the 5. and the 3. Lemma:

(e) $I/n, p, r/$ is monocal by (1) and epical by (2), hence isocal. By (3) $K/n, p, r/$ is isocal. Hence $/n, p, r + 1/$ is isocal by (4).

(f) $/n, p/$ is isocal by applying (e) inductively up to $r = \max\{p + 1, n - p + 2\}$ and (5).

(g) By definitions it is $o/H_n(X) = /n, o/$. Hence by (f) and (6) $p/H_n(X)$ is isocal for each $p \geq 0$ inductively. We may stop the calculation at $p = n$, because we have by the exact sequence of the pair (X, X^n) the epimorphism uv.

$$\to H_n(X^n) \underset{v}{\to} H_n(n/X) \underset{u}{\to} H_n(X) \to H_n(X, X^n) = 0 \to \ .$$

Hence u is an epimorphism, and therefore $n/H_n(X) = H_n(X)$.

(h) The identity (5) above, $/n, p, r/ \overset{=}{\to} /n, p/$, if $r \geq \max\{p + 1, n - p + 2\}$, is an immediate consequence of the definition

$$/n, p, r/ = \text{image}(H_n(p/X, p - r/X) \to H_n(p + r - 1/X, p - 1/X)) \ .$$

For it is $p - r/X = \emptyset$, if $r \geq p + 1$, by definition of the filtration p/X, and it is $(p + r - 1/X)_i = X_i$ up to dimension $i = n + 1$, if $r \geq n - p + 2$, because of $(Y^{p+r-1})_i = Y_i$ up to $i = n + 1$ for $p + r - 1 \geq n + 1$.

Remark: Clearly the identity (5) is true also for the spectral sequences $_i/n, p, r/$ and causes no calculation steps. Likewise the epicalculation for (2) requires no calculation step in reality: the factorization $_i/n + 1, p + r, r/ \to I_i/n, p, r/ \to I/n, p, r/$ implies that an epimorph index i for $/n + 1, p + r, r/$ can be taken as epimorph index for $I/n, p, r/$. We do not point out all simplifications, which should be made in case of an implementation. They chiefly result from the fact, that one has to do with very special cases of the Five-Lemma Algorithm, or, that additional information is available. For example in our case above the approximating chain complexes are exact themselves, which need not be in general. See the last section of the homotopy group note [2].

Final Remarks

1. In case the colimit map $f : X \to Y$ (of the proposition above) is a fibration (with fiber F and simple coefficients $H_i(F)$, $i \geq 0$), then the extended calculability of the homology of the base and of the fiber implies the extended calculability of the $/n, p, 2/$-terms. This is shown in [7], and hence by the proposition above we have the extended calculability of the homology of the total space. Furthermore the extended calculability of the base (fiber) can be deduced from the one of the fiber (base) and the total space. Further specializations yield then the calculability of the homology groups of Postnikov-complexes and of homotopy groups of simplicial complexes. Details are given in [2], [7], [10].

2. The proposition above can be interpreted as a consequence of a generalized Serre-class theory: we consider classes of sequences of abelian groups $G_i \to G_{i+1}$, $i \geq 1$. A generalized Serre-class CL is defined to be the intersection $EP \cap MO$ of two such classes (the "epiclass" EP, and the "monoclass" MO) having the obvious properties, which can be abstracted from the proposition above. In our

case CL, EP, and MO are then the classes of the isocal, epical resp. monocal, precalculable objects in AB. cf. [8, p.504].

3. No practical application (= actual calculation) has taken place up to now. Because our algorithms are not primitive recursive, there is no simple counting of steps, which could be the basis for complexity or efficiency considerations. The efficiency problems suggested in the last section of [2] amount to improving parts of the algorithms, so that the number of steps is minimized.

4. In applying the term "effective" we take the Computer Algebra view of [9]. In particular, "effective algebraic topology" should not be confused with "constructive algebraic topology" in the sense of [11, p.26], where **every** existence of objects is meant constructively (for example it would not be allowed to speak of colimits of sequences of groups as we did, and in such a theory two maps are homotopic by definition, if the one can be deformed into the other by a constructible homotopy). [13] is a thorough treatment of the various forms of constructive mathematics.

References

[1]   M.K. Agoston, Algebraic Topology, A First Course, Marcel Dekker, Inc., New York and Basel 1967
[2]   R. Schön, An algorithm for calculating homotopy groups, Memoirs of the AMS, this issue
[3]   K. Lamotke, Semisimpliziale algebraische Topologie, Die Grundlehren der math. Wissenschaften 147, Springer-Verlag 1968
[4]   S. Mac Lane, Homology, Die Grundlehren der math. Wissenschaften 114, Springer-Verlag 1967
[5]   T.B. Pinkerton, An algorithm for the automatic computation of integral homology groups, Math. Algorithms 1 (1966) 27–44
[6]   P. Ribenboim, Rings and Moduls, Interscience Publishers, New York 1969
[7]   R. Schön, Fibrations with Calculable Homology, Memoirs of the AMS, this issue
[8]   E.H. Spanier, Algebraic Topology, McGraw-Hill Book Company, New York 1966
[9]   J.H. Davenport, Effective Mathematics – the Computer Algebra viewpoint, Lecture Notes in Mathematics 873, 31–43, Springer 1981 (Proceedings, New Mexico 1980, title: Constructive mathematics)
[10]   R. Schön, The effective computability of k-invariants, Memoirs of the AMS, this issue
[11]   M.J. Beeson, Foundations of Constructive Mathematics, Springer, Berlin 1984
[12]   J. Mc Cleary, User's Guide to Spectral Sequences, Publish or Perish, Inc., Wilmington, Delaware (U.S.A.) 1985
[13]   A.S. Troelstra, D. van Dalen, Constructivism in Mathematics, North-Holland, Amsterdam 1988

Friedensstr. 2
6900 Heidelberg
Fed. Rep. of Germany

Fibrations with Calculable Homology

Abstract. Filtering the total complex X of a semisimplicial map $f : X \longrightarrow Y$ by the preimages of the skeletons of Y we get a spectral sequence

$$/n, p, r/ = {}_n E_p^r(X) \, ,$$

which converges to the homology groups of X. In a previous paper it was shown, that in case f is (algorithmically) the colimit of maps between (dimensionwise) finite complexes, then the extended calculability of the $/n, p, 2/$-terms implies the extended calculability of these groups. This paper shows the both other calculation possibilities one would like to have, namely: having calculated certain homology groups $H_k(X)$ and $/n, p, 2/$-terms for $p \leq m$ (resp. $n - p \leq m$), one can calculate the next base term $/m + 1, m + 1, 2/$ (resp. the next fiber term $/m + 1, 0, 2/$). As a consequence we have this fundamental calculation theorem for fiber spaces: each homology group of one of the complexes F, X, Y of a fibration $F \longrightarrow X \longrightarrow Y$ is extended calculable, if each homology group of the two other complexes is extended calculable. Applications are given to loop spaces, and rational homotopy groups.

Introduction

This paper is a continuation of [9], which is assumed to be known. Especially we use (without any further comment) the Five-Lemma algorithm, how it enters the proofs, and that its suppositions are fulfilled (which is obviously true in all cases which occur); cf. the proof of the proposition [9, p.8] for that. All occurring groups are assumed to be, or turn out to be (by their Serre-class property) finitely generated and abelian, which will not be mentioned any more. The spectral sequence terms $/n, p, r/$ and the differentials $d_r: /n, p, r/ \longrightarrow /n - 1, p - r, r/$ have been defined in [9, p.6]. The notations $I/n, p, r/, K/n, p, r/$ have been introduced for the image resp. the kernel of the differential going into resp. from $/n, p, r/$. Additionally we use (for cokernel and coimage) the notations $coK/n, p, r/$ and $coI/n, p, r/$, which are the quotients of $/n, p, r/$ by $I/n, p, r/$ resp. $K/n, p, r/$. They can also be used for defining homology: the fundamental property of spectral sequences (4) of [9, p.8] can be rewritten

$$0 \longrightarrow /n, p, r + 1/ \longrightarrow coK/n, p, r/ \longrightarrow coI/n, p, r/ \longrightarrow 0 \, ,$$

and the short exact filtration sequence (0) of [9, p.7] can be rewritten

$$0 \longrightarrow /n, p/ \longrightarrow H_n(X)_{p-1} \longrightarrow H_n(X)_p \longrightarrow 0 \, ,$$

where $H_n(X)_p$ denotes the quotient $H_n(X)/(p/H_n(X))$. We recall $p/H_n(X) =$ image $(H_n(p/X) \longrightarrow H_n(X))$ with p/X this part of X, which lies over the p-skeleton $Y^p \subset Y$.

1980 Mathematics Subject Classification (1985 Revision).
Primary 55 T 10, Secondary 55-04, 55 R 20, 55 R 05

Let us furthermore mention the following two facts, which are true generally (i.e. not only for fibrations), and are immediate consequences of the definition

$$/n,p,r/ = \text{image } (H_n(p/X, p - r/X) \to H_n(p + r - 1/X, p - 1/X)) \,.$$

Fact 1: $/n,p,r/ = 0$, if $p < 0$ (which is trivial, because $X^p \neq \emptyset$ for $p < 0$, and hence $p/X = \emptyset$ for $p < 0$)

Fact 2: $/n,p,r/ = 0$, if $n - p < 0$. Proof: With $Y = p + r - 1/X$ we have the exact sequence of the triple $(Y, p - 1/X, X^n)$

$$\to 0 = H_n(Y, X^n) \to H_n(Y, p - 1/X) \to H_{n-1}(p - 1/X, X^n) = 0 \to$$

and hence $H_n(Y, p - 1/X) = 0$.

By reasons of completeness and uniformity let us reformulate the essential part of the proposition [9, p.8] as follows.

Proposition (for Calculating the Total Complex:)

If for a simplicial map $f : X \to Y$ each $/n,p,2/$-term is extended calculable, then each homology group $H_n(X)$ is extended calculable.

The results of this paper now read as follows:

Proposition (for Calculating the Next Base Term):

Let $m \geq 0$ be a given fixed integer. If

(i) each term $/n,p,2/$, $p \leq m$, is extended calculable, and
(ii) the homology groups $H_m(X)$ and $H_{m+1}(X)$ are extended calculable,

then the base term $/m + 1, m + 1, 2/$ is extended calculable.

Proposition (for Calculating the Next Fiber Term):

Let $m \geq 0$ be a given fixed integer. If

(i) each term $/n,p,2/$, $n - p \leq m$, is extended calculable, and
(ii) the homology groups $H_{m+1}(X)$ and $H_{m+2}(X)$ are extended calculable,

then the fiber term $/m + 1, 0, 2/$ is extended calculable.

The proofs are given in section **I** resp. **II**. The following theorem is proved in **III**. Applications are given in **IV**.

Theorem (for Calculating Homology of Fiber Spaces):

Let $F \to X \to Y$ be a fibration with F, Y connected Kan-complexes, and simple coefficients $H_i(F)$, $i \geq 0$. If each homology group of two of the complexes F, X, Y is extended calculable, then each one of the third complex is extended calculable.

Here also we have left out the (in our case self-evident) assumption, that $X \to Y$ is the colimit of a preconstructible sequence of maps $_i f : {}_i X \to {}_i Y$ between finite complexes, see (b) of [9, p.8] and let us consider from now on the following agreements as self-evident, too:

(a) The complexes $_iY$ and Y have distinguished base-point complexes pt, the maps $_iY \to {}_{i+1}Y \to Y$ are base-point preserving, and the fibers $_iF$ (over pt in $_iY$) form the approximating spaces for the fiber F (over pt in Y).

(b) The induced map over an $\underline{Y} = {}_iY$ (by the map $_iY \to Y$ has as approximating maps the induced maps over $\underline{Y}$ by the maps $\underline{Y} \to Y$, $k \geq i$, from the maps $_kX \to {}_kY$ (i.e. we keep the base complex $\underline{Y}$ stationary).

(c) If $u : A \to \underline{Y}$ is an arbitrary map with A (dimensionwise) finite and $\underline{Y}$ as in (b), the induced map over A (by $A \to \underline{Y} \to Y$ from $X \to Y$) has as approximating maps the induced maps over A by $A \to \underline{Y} \to {}_kY$, $k \geq i$, from $_kX \to {}_kY$.

See the well-known switching lemma of colimits with limits in the category of sets, e.g. in [11, p.73], for the well-definedness of (a), (b) and (c).

(d) We need another construction later on: the **first Eilenberg-subcomplex** ^1X of an (arbitrary) complex X with base point; its n-simplexes are these $x \in X_n$, which map the 0-skeleton of $\Delta[n]$, the standard n-simplex, onto the base point. The inclusion $^1X \to X$ is a homotopy equivalence, if X is a connected Kan-complex. If X has the approximating complexes $_iX$, then ^1X has the approximating complexes $^1{}_iX$, and it is clear, that an Eilenberg-subcomplex $Y = {}^1X$ fulfills the condition $^1Y = Y$.

(e) With $IP = \Delta[1] \times P$, $P = \Delta[p]$, we have the **canonical homotopy** 5.4 of [5, p.12] $H : IP \to P$, and its restriction $\dot{H} : I\dot{P} \to P$, $\dot{P}$ the $(p-1)$-skeleton of P, which go from the constant map onto (0) to the identity $P \to P$ resp. to the inclusion $\dot{P} \to P$. The canonical bottom and top inclusions $\underline{i} : P \to IP$, $\bar{i} : \dot{P} \to I\dot{P}$, $i = 0,1$, induce isomorphisms in homology between the total spaces of the fibrations, which are induced from $X \to Y$ by the maps yH and $yH\underline{i}$ resp. $y\dot{H}$ and $y\dot{H}\bar{i}$, $y : P \to Y$ a p-simplex: the maps lying over $\underline{i}, \bar{i}$ are denoted by

$$Y_{(i)/P} \to Y_{/IP} \quad \text{resp.} \quad Y_{(i)/\dot{P}} \to Y_{/I\dot{P}} , \quad i = 0,1 ,$$

where $/A$ denotes the appertaining base space A.

(f) By definitions it is $(Y_{(1)/P}, Y_{(1)/\dot{P}}) = (P \times_Y X, \dot{P} \times_Y X)$, and if the **additional assumption** $Y = {}^1Y$ is fulfilled, it is moreover $(Y_{(0)/P}, Y_{(0)/\dot{P}}) = (P \times F, \dot{P} \times F)$. By the classical Five Lemma we have the isomorphisms $0', 1'$:

(g)

$$H_n(P \times F, \dot{P} \times F) \xrightarrow[\cong]{0'} H_n(Y_{/IP}, Y_{/I\dot{P}}) \xleftarrow[\cong]{1'} H_n(P \times_y X, \dot{P} \times_y X)$$

$$\cong \uparrow \qquad\qquad\qquad\qquad\qquad\qquad\qquad \uparrow =$$

$$H_{n-p}(F) \qquad\qquad\qquad\qquad\qquad\qquad\qquad {}_yP/n,p,1/$$

The first (natural, vertical) isomorphism is given by the cross-product with an element of $H_p(P, \dot{P})$. The last identity is given by definition of $_yP/n, p, 1/$, the $/n, p, 1/$-term associated to the fibration with base P, induced from $X \to Y$ by y.

I Calculating the Base Term

1. Lemma

If each term $/n, p, 2/$, $p \leq m$, is isocal, then each term

$/n, p, r/$ is isocal for $p \leq m - r + 1$, $(a)_r$

$/n, p, r/$ is epical for $p \leq m$, $(b)_r$

$/n, p/$ is epical for $p \leq m$, (c)

$coK/n, p, r/$ is epical for $p \leq m$, (d)

$coK/n, p, r/$ is isocal for $p \leq m - r$, (e)

$coI/n, p, r/$ is monocal for $p \leq m + 1$. (f)

Proof

Consider the exact sequences
(1) $0 \to /n, p, r/ \xrightarrow{=} /n, p/ \to 0$, if $r \geq \max\{p + 1, n - p + 2\}$ ·
(2) $/n + 1, p + r, r/ \to /n, p, r/ \to coK/n, p, r/ \to 0$
(3) $0 \to coI/n, p, r/ \to /n - 1, p - r, r/$
(4) $0 \to /n, p, r + 1/ \to coK/n, p, r/ \to coI/n, p, r/ \to 0$
and the following facts, which are immediate:
(C) $(b)_r$, (1) imply (c), if $r = \max\{p + 1, n - p + 2\}$.
(D) $(b)_r$, (2) imply (d).
(E) $(a)_r$, $(b)_r$, (2) imply (e), because by $(b)_r$ $/n + 1, p + r, r/$ is epical for $p + r \leq m$,
   i.e. for $p \leq m - r$.
(F) $(a)_r$, (3) imply (f), because by $(a)_r$ $/n - 1, p - r, r/$ is isocal for $p - r \leq m - r + 1$,
   i.e. for $p \leq m + 1$.

Hence it suffices to prove $(a)_r$, $(b)_r$ by induction on $r \geq 2$. $(a)_2$, $(b)_2$ are true, because
$/n, p, 2/$ is isocal for $p \leq m$ by supposition.
   Let $(a)_r$, $(b)_r$ be true, $r \geq 2$ and consider (4). $coK/n, p, r/$ is isocal for
$p \leq m - r = m - (r + 1) + 1$ by (e), and epical for $p \leq m$ by (d). $coI/n, p, r/$
is monocal for $p \leq m + 1$ by (f). Hence by (4) $(a)_{r+1}$ resp. $(b)_{r+1}$ is true.

2. Lemma

Let $m, n \geq 0$ be given integers. If
(i) $H_n(X)$ is isocal, and
(ii) $/n, p/$ is epical for $0 \leq p \leq m$, then for each $p \leq m$ it is true

$(a)_p : H_n(X)_{p-1}$ is isocal,   and

$(b)_p : /n, p/$ is isocal.

   If $n = m + 1$, then for each p

$(c) : /n, p/$ is isocal.

Proof

Consider the exact sequence

$$(5)_p : \quad 0 \to /n, p/ \to H_n(X)_{p-1} \to H_n(X)_p \to 0$$

$(a)_o$ is true, because $H_n(X)_{-1} = H_n(X)$ is isocal by supposition. $(b)_o$ is true, because $/n, 0/$ is monocal by (i) and $(5)_o$, and epical by (ii). Let $(a)_p, (b)_p$ be true for a p with $0 \leq p < m$. Then $H_n(X)_p$ is isocal by $(5)_{p+1}$ which is $(a)_p$, and hence by $(5)_p/n, p+1/$ is monocal. Together with (ii) we have $/n, p+1/$ is isocal, which is $(b)_{p+1}$. Proof of (c): Let be $n = m + 1$, and consider the exact sequences

$$(6)_p : \quad 0 \to (p-1)/H_n(X) \to p/H_n(X) \to /n, p/ \to 0 , \quad \text{and}$$

$$(7) : \quad 0 \to (m-1)/H_n(X) \to H_n(X) \to H_n(X)_{m-1} \to 0 .$$

By $(a)_m$, (i) and (7) $(m-1)/H_n(X)$ is isocal. Hence with $(b)_m$ and $(6)_m$ $m/H_n(X)$ is isocal. Therefore with $(6)_n$, the identity $n/H_n(X) = H_n(X)$, see (g) of [9, p.9], and (i): $/n, n/$ is isocal. Together with $(b)_p$ for $p < n$, and $/n, p/ = 0$ for $p > n$ (c) is proved.

3. Lemma

Let $m \geq 0$, $q \geq 0$ be given integers. If

(i) each term $/n, p, 2/$, $p \leq m$, is isocal, and
(ii) the term $/m, m - q/$ is isocal,
then for $r > q \geq 0$, this statement is true:

    (r) : the term $coK/m, m - q, r/$ is monocal.

Proof

At first an observation: if $r > q$, then by Fact 2 of the introduction it is $/m + 1, m - q + r + 1, r + 1/ = 0$, and therefore we have

    (iii) $coK/m, m - q, r + 1/ = /m, m - q, r + 1/ , \quad \text{if} \quad r > q .$

Now let be $r \geq R = \max\{m - q + 1, q + 2\}$. Then it is $r - 1 > q$ and by **I** $coK/m, m - q, r/ = /m, m - q, r/ = /m, m - q/$ (the latter identity by (1) of the 1. Lemma), terms, which are isocal by (ii). Let $(r + 1)$ be true for an r with $q < r < R$, and consider the exact sequence

$$0 \to /m, m - q, r + 1/ \to coK/m, m - q, r/ \to coI/m, m - q, r/ \to 0 ,$$

which is a special case of (4) of the 1. Lemma. By (iii) and the induction hypothesis $/m, m - q, r + 1/$ is monocal. By (i) we can apply (f) of the 1. Lemma, and have $coI/m, m - q, r/$ is monocal (because $m - q \leq m + 1$). Therefore the middle term $coK/m, m - q, r/$ is monocal, which finishes the induction step.

Proposition (for Calculating the Next Base Term)

Let $m \geq 0$ be a given fixed integer. If

(i) each term $/n, p, 2/, p \leq m$, is extended calculable, and
(ii) the homology groups $H_m(X)$ and $H_{m+1}(X)$ are extended calculable,

then the base term $/m + 1, m + 1, 2/$ is extended calculable.

Proof

We show, that for all $r \geq 2$, this statement is true:

(r) the term $/m + 1, m + 1, r/$ is isocal.

If $r \geq R = \max\{m+2, 2\} = m+2$, then $/m+1, m+1, r/ = /m+1, m+1/$ is isocal by (c) of the 2. Lemma, because its conditions H_{m+1} isocal, and $/n, p/$ epical for $p \leq m$ (see (c) of the 1. Lemma) are fulfilled. Let $(r + 1)$ be true for an r with $2 \leq r < R$, and consider the exact sequences

$$0 \to K/m + 1, m + 1, r/ \underset{u}{\to} /m + 1, m + 1, r + 1/ \to 0 , \quad \text{and}$$

$$0 \to K/m + 1, m + 1, r/ \to /m + 1, m + 1, r/ \overset{d}{\to} /m, m - r + 1, r/$$
$$\downarrow$$
$$coK/m, m - r + 1, r/ \to 0 .$$

The isomorphism u results from Fact 2 of the introduction, because $/m + 2, m + 1 + r, r/ = 0$ implies, that $I/m + 1, m + 1, r/ = 0$, and hence $/m + 1, m + 1, r + 1/ \cong K/m + 1, m + 1, r/$ by the fundamental property of spectral sequences. The second exact sequence is a standard elementary fact from homological algebra. By induction hypothesis, we may assume, that $/m+1, m+1, r+1/$ and hence (via u), that $K/m + 1, m + 1, r/$ is isocal. $/m, m - r + 1, r/$ is isocal by $(a)_r$ of the 1. Lemma with $p = m - r + 1$, and (i) above. By (c) of the 1. Lemma $/n, p/$ is epical for $0 \leq p \leq m$. With $(b)_{m-q}, q = r - 1$, of the 2. Lemma (with $n = m$) $/m, m - q/$ is isocal. Hence by the 3. Lemma $coK/m, m - r + 1, r/$ is monocal. Therefore the term $/m + 1, m + 1, r/$ is isocal, which finishes the induction step.

Remark

If $f : X \to Y$ is a fibration with connected fiber F, then the term $/m + 1, m + 1, 2/$ above is extended calculable if and only if $H_{m+1}(Y)$ is extended calculable. For the proof see section **III**.

II Calculating the Fiber Term

4. Lemma   (Notation: $q = n - p$)

If each term $/n, p, 2/, q \leq m$, is isocal, then each term

$/n, p, r/$ is isocal for $q \leq m - r + 2$, $(a)_r$

$/n, p, r/$ is monocal for $q \leq m$, $(b)_r$

$/n, p/$ is monocal for $q \leq m$, (c)

$K/n, p, r/$ is monocal for $q \leq m$, (d)

$K/n, p, r/$ is isocal for $q \leq m - r + 1$, (e)

$I/n, p, r/$ is epical for $q \leq m + 1$. (f)

Proof

Consider the exact sequences

(1) $0 \to /n, p, r/ \xrightarrow{=} /n, p/ \to 0$, if $r \geq \max\{p + 1, q + 2\}$

(2) $0 \to K/n, p, r/ \to /n, p, r/ \to /n - 1, p - r, r/$

(3) $/n + 1, p + r, r/ \to I/n, p, r/ \to 0$

(4) $0 \to I/n, p, r/ \to K/n, p, r/ \to /n, p, r + 1/ \to 0$

and the following facts, which are immediate

(C) $(b)_r, (1)$ imply (c), if $r = \max\{p + 1, n - p + 2\}$.

(D) $(b)_r, (2)$ imply (d)

(E) $(a)_r, (b)_r, (2)$ imply (e) because by $(b)_r /n - 1, p - r, r/$ is monocal for

   $(n - 1) - (p - r) \leq m$, i.e. for $q = n - p \leq m - r + 1$.

(F) $(a)_r, (3)$ imply (f), because by $(a)_r /n + 1, p + r, r/$ is isocal for

   $(n + 1) - (p + r) \leq m - r + 2$, i.e. for $q = n - p \leq m + 1$.

Hence it suffices to prove $(a)_r, (b)_r$ by induction on $r \geq 2$. $(a)_2, (b)_2$ are true, because $/n, p, 2/$ is isocal for $q \leq m$ by supposition. Let $(a)_r, (b)_r$ be true, $r \geq 2$, and consider (4). $K/n, p, r/$ is isocal for $q \leq m - r + 1 = m - (r + 1) + 2$ by (e), and monocal for $q \leq m$ by (d). $I/n, p, r/$ is epical for $q \leq m + 1$ by (f). Hence by (4) $(a)_{r+1}$ resp. $(b)_{r+1}$ is true.

5. Lemma

Let $m, n \geq 0$ be given integers. If

 (i) $H_n(X)$ is isocal, and

(ii) $/n, n - q/$ is monocal for $0 \leq q \leq m$,

then for each $p \geq n - m$ it is true

   $(a)_p : p/H_n(X)$ is isocal, and

   $(b)_p : /n, p/$ is isocal.

If $n = m + 1$, then for each p

   $(c) : /n, p/$ is isocal.

Proof

Consider the exact sequence

$$(5)_p : 0 \to (p-1)/H_n(X) \to p/H_n(X) \to /n,p/ \to 0 .$$

For $p \geq n+1$ it is $p/H_n(X) = (p-1)/H_n(X) = H_n(X)$, and it follows, that $(a)_p$ is true by (i), and that $/n,p/ = 0$, i.e. $(b)_p$ is true. Let $(a)_p, (b)_p$ be true for a p with $n-m < p \leq n+1$. Then by $(5)_p$ $(a)_{p-1}$ is true. By $(5)_{p-1}$ $/n,p-1/$ is therefore epical, and hence with (ii) isocal, because with $q = n-p+1$ it is $p-1 = n-q$ and $0 \leq q \leq m$. Therefore $(b)_{p-1}$ is true, which finishes the induction step. Proof of (c): Let be $n = m+1$. Consider the exact sequences

$$(6)_p : 0 \to p/H_n(X) \to H_n(X) \to H_n(X)_p \to 0 \quad \text{and}$$

$$(7) : 0 \to /n,1/ \to H_n(X)_o \to H_n(X)_1 \to 0 .$$

By $(a)_1$, (i), and $(6)_1$ $H_n(X)_1$ is isocal. Hence with $(b)_1$ and (7) $H_n(X)_o$ is isocal. Therefore with $(6)_o$ and (i) $0/H_n(X) = /n,o/$ is isocal. Together with $(b)_p$ for $p \geq n-m = 1$, and $/n,p/ = 0$ for $p < 0$ (c) is proved.

6. Lemma   (Notation: $q = n - p$)

Let $m \geq 0$, $\underline{p} \geq 2$ be given integers. If each term $/n,p,2/$, $q \leq m$, is isocal, and the term $/m+2,\underline{p}/$ is isocal, then for $r \geq \underline{p} \geq 2$ this statement is true:

$$(r) : K/m+2,\underline{p},r/ \quad \text{is epical.}$$

Proof

For $r \geq R = \max\{\underline{p}+1, m+2-\underline{p}+2\}$ it is $r > \underline{p}$, and therefore $/m+1,\underline{p}-r,r/ = 0$ by Fact 1 of the introduction. Hence $K/m+2,\underline{p},r/ = /m+2,\underline{p},r/ = /m+2,\underline{p}/$ are terms, which are isocal by supposition. Let $(r+1)$ be true for an r with $1 \leq \underline{p} \leq r < R$; then it is $r+1 > \underline{p}$, and again we have $K/m+2,\underline{p},r+1/ = /m+2,\underline{p},r+1/$, a term, which is epical by induction hypothesis. It is the quotient term in the exact sequence

$$0 \to I/m+2,\underline{p},r/ \to K/m+2,\underline{p},r/ \to /m+2,\underline{p},r+1/ \to 0 .$$

By (f) of the 4. Lemma $I/m+2,\underline{p},r/$ is epical, because $m+2-\underline{p} \leq m+1$. Hence the middle term $K/m+2,\underline{p},r/$ is epical, which finishes the induction step.

Proposition (for Calculating the Next Fiber Term)

Let $m \geq 0$ be a given fixed integer. If

(i) each term $/n,p,2/$, $n-p \leq m$ is extended calculable, and
(ii) the homology groups $H_{m+1}(X)$ and $H_{m+2}(X)$ are extended calculable,
then the fiber term $/m+1,0,2/$ is extended calculable.

Proof

We show, that for all $r \geq 2$, this statement is true:

(r)   the term $/m+1,0,r/$ is isocal.

If $r \geq R = \max\{1, m+1+2\} = m+3$, it is $/m+1,0,r/ = /m+1,0/$, a term, which is isocal by (c) of the 5. Lemma, because its conditions (with $n = m+1$) (i) $H_{m+1}(X)$ is isocal, and (ii) $/m+1, m+1-q/$ is monocal for $0 \leq q \leq m$ (see (c) of the 4. Lemma) are fulfilled.

Let $(r+1)$ be true for an r with $2 \leq r < R$, and consider the exact sequence

$$0 \to /m+1,0,r+1/ \xrightarrow{u} coK/m+1,0,r/ \to coI/m+1,0,r/ \to 0 \ .$$

It is $coI/m+1,0,r/ = 0$ by Fact 1 of the introduction, because $/m,-r,r/ = 0$ implies, that $K/m+1,0,r/ = /m+1,0,r/$; hence u is an isomorphism. Consider now

$$0 \to K/m+2,r,r/ \to /m+2,r,r/ \xrightarrow{d} /m+1,0,r/ \to coK/m+1,0,r/ \to 0 \ .$$

By induction hypothesis $/m+1,0,r+1/$ and hence (via the isomorphism u) $coK/m+1,0,r/$ is isocal. By (c) of the 4. Lemma $/m+2, m+2-q/$ is monocal for $0 \leq q \leq m$. With $(b)_p$ of the 5. Lemma (with $n = m+2$) $/m+2,p/$ is isocal for each $p \geq n-m = 2$. Hence by the 6. Lemma with $\underline{p} = r \geq 2 : K/m+2,r,r/$ is epical. $/m+2,r,r/$ is isocal by $(a)_r$ of the 4. Lemma with $p = r$ and $n = m+2$ (regard, that then $q = n-p = m+2-r \leq m-r+2$). Hence the term $/m+1,0,r/$ is isocal, which finishes the induction step.

Remarks

1. If $f : X \to Y$ is a fibration with connected base Y and simple coefficients $H_i(F)$, $i \geq 0$, F the fiber, then the term $/m+1,0,2/$ above is extended calculable, if and only if $H_{m+1}(F)$ is extended calculable. For the proof see section **III**.

2. The proof of section **II** is obviously analogous to the one of **I**, and there exists an appropriate categorial framework, where these proofs coincide, but which is less practical algorithmically.

3. The used exact sequences are no surprise for those readers, who once have checked all details of the proof of the comparison theorem for spectral sequences in the form of [15, p.356], known at least since 1956 by Kudo and Araki [22]. Only the steps from the total complex to the infinity terms are added here.

   It should be noted, that the **indirect** proof of the comparison theorem by Zeeman [23] gives no appropriate diagram of exact sequences, after which our algorithm could be modelled. The same is true for other, simpler diagrams, e.g. [13, p.507], which suffice to transport the classical Serre-class properties. References for corresponding exact sequences in the cohomology case are [7, p.56], [16, p.57].

III Calculating Homology of Fiber Space

The definition of the spectral sequence terms

$$/n, p, r/X = \text{image } (H_n(p/X, (p-r)/X) \to H_n((p+r-1)/X, (p-1)/X))$$

extends to pairs as follows:

$$/n, p, r/(X, A) = \text{image } (H_n(A \cup p/X, A \cup (p-r)/X)$$
$$\to H_n(A \cup (p+r-1)/X, A \cup (p-1)/X)) .$$

It is obvious from these definitions, that a $/n, p, 1/$-term of X is identical with a certain $/n, p, 2/$-term, namely:

$$/n, p, 1/X = /n, p, 2/(p/X, (p-1)/X) ,$$

with p/X finitely filtered by $i/(p/X) = i/X$ for $i \leq p$, and $i(p/X) = p/X$ for $i \geq p$. Hence having established a natural $/n, p, 2/$-term isomorphism for pairs of fibrations, e.g. [5, p.156], we immediately get a natural isomorphism for $/n, p, 1/$-terms, namely (with $f : X \to Y$ a fibration):

$$Y/n, p, 1/ = /n, p, 1/X \cong H_p(Y^p, Y^{p-1}; H_{n-p}(F)) ,$$

the first identity is a notation, F is the fiber and $H_{n-p}(F)$ in general a local coefficient system. With help of this isomorphism and the excision property for homology, we get a direct sum decomposition

$$(1) \quad \oplus_y P/n, p, 1 \xrightarrow{\cong} Y/n, p, 1/$$

with indexing set all non-degenerate p-simplexes $y : P \to Y$ (which induce the homomorphisms $_y P/n, p, 1/ \to Y/n, p, 1/$; see (g) of the introduction for notation).

1. Proposition   (for calculating base/fiber homology from base/fiber terms)

Let $F \to X \to Y$ be a fibration with simple coefficients $H_i(F)$, $i \geq 0$. Then

(a) If F is connected, then the base term $/n, n, 2/$ is extended calculable, if and only if $H_n(Y)$ is extended calculable.

(b) If Y is connected, then the fiber term $/n, 0, 2/$ is extended calculable, if and only if $H_n(F)$ is extended calculable.

Proof

Consider the following squares (a), (b) with the obvious maps.

$$
\begin{array}{ccccc}
F & \to & X & \to & Y \\
\downarrow & (b) & \downarrow & (a) & \downarrow = \\
pt & \to & Y & \to & Y
\end{array}
$$

(a) Square (a) induces an isomorphism $H_n(Y, H_o(F)) \xrightarrow{\cong} H_n(Y, H_o(pt))$, or equivalently an isomorphism $/n, n, 2/X \xrightarrow[\cong]{} /n, n, 2/Y \xrightarrow[\cong]{} H_n(Y)$. The latter one (which is the $/n,n,2/$-term isomorphism of the identity $Y \to Y$) is natural in Y and true also for the approximation complexes $_iY$. Hence the algorithm for an isocalculation of $H_n(Y)$ can be taken as an algorithm for an isocalculation of $/n,n,2/Y$ (without knowing, how the isomorphism looks like). The application of the Five-Lemma algorithm to the first isomorphism (existing in the colimit) gives now the desired isocalculation of $/n,n,2/ = /n,n,2/X$. With the obvious reverse reasoning (a) is proved

(b) Square (b) induces an isomorphism $H_o(pt, H_n(F)) \to H_o(Y, H_n(F))$, or equivalently an isomorphism $H_n(F) = pt/n,0,2/ \underset{\cong}{\to} Y/n,0,2/ = /n,0,2/$. Hence with the Five-Lemma algorithm (b) is proved.

Remark

In case (a) assumptions on the coefficients can be dropped, because $H_o(F)$, which solely enters the proof, is always simple, if F is connected.

2. Proposition   (for calculating $/n,p,2/$-terms from base and fiber homology)

Let $F \to X \to Y$ be a fibration with simple coefficients $H_i(F)$, $i \geq 0$, and F, Y connected Kan-complexes. If $H_{p-1}(Y), H_p(Y)$, and $H_{n-p}(F)$ are extended calculable, then the term $/n,p,2/$ is extended calculable.

Proof

Because $/n,p,2/$ is known to be isomorphic to $H_p(Y, H_{n-p}(F))$, a homology group, which is (effectively) calculable by the tensor-Tor formula, it suffices by (d) of the 3. Lemma of [9] to show the calculability of an epimorph integer. By the obvious fact (5) of [9, p.3], and the isomorphism $H_i(^1Y) \to H_i(Y)$, $i = p$ or $i = p-1$, see (d) of the introduction, we may assume, that an index j is calculated, which is split epimorph for $H_{p-1}(^1Y)$ and epimorph for $H_p(^1Y)$. By the universal coefficient sequence we have then the epimorphism

$$H_p(^1_jY, H_{n-p}(F)) \to H_p(^1Y, H_{n-p}(F)) \overset{\cong}{\to} H_p(Y, H_{n-p}(F)) ,$$

and (equivalently) the epimorphism

$$^1_jY/n,p,2/ \to {}^1Y/n,p,2/ \overset{\cong}{\to} Y/n,p,2/ .$$

It suffices to calculate an epimorph integer for $^1_jY/n,p,2/$, say k, because then the epimorphism

$$^1_jY(_k/n,p,2/) \to {}^1_jY/n,p,2/ \to Y/n,p,2/ ,$$

which factorizes through $Y(_k/n,p,2/)$, implies, that k is an epimorph integer for $Y/n,p,2/$.

We show by the following lemma, that $^1_jY/n,p,1/$ is extended calculable, which implies (by the proposition (e) of [9, p.8]), that $^1_jY/n,p,2/$ is even extended calculable, and the proof is ended.

Lemma   (for calculating $/n, p, 1/$-terms from fiber homology)

Let $F \to X \to Y$ be a fibration, with Y finite and $^1Y = Y$ (see (d) of the introduction), and with stationary approximating base complexes $_iY = Y$. If $H_{n-p}(F)$ is extended calculable, then the term $Y/n, p, 1/$ is extended calculable.

Proof

Because $Y/n, p, 1/$ is known to be isomorphic to so many copies of $H_{n-p}(F)$ as there are non-degenerate p-simplexes in Y, $Y/n, p, 1/$ is clearly (effectively) calculable. Hence it suffices to calculate an epimorph integer k for $Y/n, p, 1/$ with respect to the approximating sequence $Y(_i/n, p, 1/)$, which is induced by the complexes $_iX$ approximating X. The non-degenerate p-simplexes $y : P \to Y$ induce for each i the canonical square

$$\oplus_y P/n, p, 1/ \overset{\cong}{\to} Y/n, p, 1/$$
$$\uparrow \qquad\qquad \uparrow$$
$$\oplus_y P(_i/n, p, 1/) \to Y(_i/n, p, 1/) ,$$

where the upper map is the isomorphism (1) before the 1. Proposition above. Hence such an integer k is given by the maximum of the integers $k(y)$, where $k(y)$ is an epimorph integer for $_yP/n, p, 1/$. In order to calculate such an integer $k(y)$ we consider the diagram (g) in the introduction:

$$H_n(P \times F, \dot{P} \times F) \overset{0'}{\underset{\cong}{\to}} H_n(Y_{/IP}, Y_{/I\dot{P}}) \overset{1'}{\underset{\cong}{\leftarrow}} H_n(P \times _yX, \dot{P} \times _yX)$$
$$\cong \uparrow \qquad\qquad\qquad\qquad\qquad\qquad\qquad \uparrow =$$
$$H_{n-p}(F) \qquad\qquad\qquad\qquad\qquad\qquad _yP/n, p, 1/$$

The first isomorphism (true also for the approximating complexes $_iF$) implies, that the algorithm for an isocalculation of $H_{n-p}(F)$ can be taken as algorithm for an isocalculation of $\dot{H}_n(P \times F, \dot{P} \times F)$. The Five-Lemma algorithm gives the rest.

Theorem   (for calculating homology of fiber spaces)

Let $F \to X \to Y$ be a fibration with simple coefficients $H_i(F)$, $i \geq 0$, and F, Y connected Kan-complexes, then the following is true:
If each homology group

(a) $H_i(Y)$, $H_i(F)$, $i \geq 0$, is extended calculable, then so is $H_i(X)$, $i \geq 0$,
(b) $H_i(X)$, $H_i(F)$, $i \geq 0$, is extended calculable, then so is $H_i(Y)$, $i \geq 0$,
(c) $H_i(X)$, $H_i(Y)$, $i \geq 0$, is extended calculable, then so is $H_i(F)$, $i \geq 0$.

Proof

(a) By the 2. Proposition each $/n, p, 2/$-term is extended calculable. Hence by (g) of [9, p.8] Proposition, each $H_i(X)$, $i \geq 0$, is extended calculable.

(b) By (b) of the 1. Proposition each fiber term $/n, 0, 2/$ is extended calculable, and hence the conditions (i) and (ii) of the Proposition (for calculating the next base term, p.16) are fulfilled for $m = 0$. Let us assume, that these conditions are fulfilled up to $m = i - 1$, $i \geq 1$. We have then the extended calculability of the base terms $/m, m, 2/$, $1 \leq m \leq i$, and also of $/0, 0, 2/$, because $/0, 0, 2/$ is a fiber term and we can use (b) of the 1. Proposition. With (a) of the 1. Proposition the **homology groups** $H_m(Y)$ **are extended calculable** up to $m = i$, and hence by the 2. Proposition the $/n, p, 2/$-terms are extended calculable up to $p = i$. Result: conditions (i) and (ii) of the Proposition (for calculating the next base term, p.15) are fulfilled up to $m = i$, and the induction step is accomplished.

(c) By (a) of the 1. Proposition each base term $/n, n, 2/$ is extended calculable, and hence the conditions (i) and (ii) of the Proposition (for calculating the next fiber term, p.18) are fulfilled for $m = 0$. Let us assume, that these conditions are fulfilled up to $m = i - 1$, $i \geq 1$. We have then the extended calculability of the fiber terms $/m, 0, 2/$, $1 \leq m \leq i$, and also of $/0, 0, 2/$, because $/0, 0, 2/$ is a base term and we can use (a) of the 1. Proposition. With (b) of the 1. Proposition the **homology groups** $H_m(F)$ **are extended calculable** up to $m = i$, and hence·by the 2. Proposition the $/n, p, 2/$-terms are extended calculable up to $n - p = i$. Result: conditions (i) and (ii) of the Proposition (for calculating the next fiber term, p.18) are fulfilled up to $m = i$, and the induction step is accomplished.

Remarks

1. The proof (and the algorithm) of the 2. Proposition considerably simplifies, if the approximating maps $_iX \to {}_iY$ are fibrations themselves: an epimorph integer for the $/n, p, 2/$-term is then immediately given by $k = \max\{r, s, t\}$ with r, s split epimorph indexes for $H_{p-1}(Y)$ resp. $H_{n-p}(F)$, and t epimorph for $H_p(Y)$. By the universal coefficient theorem we have namely then the epimorphisms

$$H_p({}_kY, H_{n-p}({}_kF)) \to H_p({}_kY, H_{n-p}(F)) \to H_p(Y, H_{n-p}(F)) \ ,$$

or equivalently the epimorphism $_k/n, p, 2/ \to /n, p, 2/$.

2. Case (a) applies also to a projection $F \times Y \to Y$ (which is no fibration, if F is no Kan-complex). If the approximating maps are projections $_iF \times {}_iY \to {}_iY$ too, then a more direct calculation is possible: $H_i(F \times Y)$ is (effectively) calculable by the tensor-Tor formula resulting from the short exact Künneth sequence, e.g. [13, p.235], which implies, that as an epimorph integer k of $H_i(F \times Y)$ we may choose the maximum of the split epimorph indexes of the homology groups $H_p(Y)$, $H_p(F)$, $0 \leq p \leq i$, given by isocalculations of these groups. (In the general case it is clear, that if $_iX$ approximates $E \times Y$, then the image complexes under the maps $_iX \to F \times Y \to Y$, $_iX \to F \times Y \to F$, form approximations of Y resp. F.)

IV Applications

Let us recall the simplicial loop-complex LX of an (arbitrary, pointed) complex X, e.g. [5, p.196]. It is the fiber of the path-map

$$p = d_o : PX \to X \ , \quad \text{with } (PX)_n = \{x/x \in X_{n+1} \ , \ x(0) = pt\} \ ,$$

with face and degeneracy operators $d_i^P = d_{i+1}$, $s_i^P = s_{i+1}$.

If X is a connected Kan-complex, then PX is contractible, and $PX \to X$ is a fibration. If $_iX$ are the approximating complexes for X, then the maps $P_iX \to {}_iX$ approximate $PX \to X$.

Proposition

Let X be a connected Kan-complex, such that each homology group $H_i(X)$ is extended calculable. If X is simply connected, then

a) each homology group $H_i(LX)$, $i \geq 0$, is extended calculable,
b) each homotopy group $\pi_i(X) \otimes \mathbb{Q}$, $i \geq 1$ (hence the free summand of $\pi_i(X)$) is effectively calculable.

Proof

(a) Consider the fibration $LX \to PX \to X$, and apply (c) of the Theorem above. Clearly the algorithm simplifies in this case because of the known vanishing of the infinity terms, which makes superfluous the 5. Lemma of section **II**.
(b) The rational homotopy group $\pi_{i+1}(X) \otimes \mathbb{Q}$ is known to be isomorphic to the group of primitive elements in $H_i(LX, Q)$, i.e. to the kernel of the map

$$D_* - \underline{1}_* - \underline{2}_* : H_i(LX, \mathbb{Q}) \to H_i(LX \times LX, \mathbb{Q}) , \quad [1, p.132] ,$$

with $D, \underline{1}, \underline{2} : LX \to LX \times LX$ the diagonal, resp. the inclusion into the first resp. second factor. By (a) $H_i(LX, \mathbb{Q})$ is extended calculable. Hence with the second remark of section **III** $H_i(LX \times LX, \mathbb{Q})$ is extended calculable. The Five-Lemma algorithm now applied to the obvious exact sequence yields then the extended calculability of kernel $(D_* - \underline{1}_* - \underline{2}_*) \cong \pi_{i+1}(X) \otimes \mathbb{Q}$ cf. [3], [17].

Remarks

1. It is shown in [10], that for a complex A of finite type (i.e. with finitely many non-degenerate simplexes) there exists a preconstructible sequence of complexes, [9, p.8], of finite type $_iA$, $i \geq 1$ such that
   (i) $_{i+1}A$ is constructed out of $_iA$ by filling a horn [2, p.68]
   (ii) the inclusion $A \to \lim({}_iA) = X$ is a homotopy equivalence into a Kan-envelope X of $A = {}_1A$.
   (iii) Each homology group $H_i(X)$ is extended calculable with respect of these approximating complexes $_iA$.
   Hence by the proposition above we have for such complexes A (with additionally assuming, that X is simply connected, i.e. for example, if A arises from a simply connected geometric simplicial complex) the effective calculability of its rational homotopy group.
2. Part a) of the proposition above cannot be iterated, because the simple connectedness gets lost. But it is shown in [10], that the second Eilenberg subcomplex $^2(LX)$ of LX has extended calculable homology, if X has. Here the extended calculability of $K(G, 1)$ complexes is needed and will be proved. We get then the effective calculability of the homotopy groups of simply connected complexes of finity type by the standard inductive reasoning.

3. There exists an extensive literature concerning or being directed towards computations of homology of loop spaces, rational homotopy groups, or homology of fiber spaces. But computations there generally (exceptions are [24] and parts of [17], [20], [21]) do not mean algorithmic computations, but rather deductions of isomorphisms $G \cong G'$ (e.g. the E_2-term isomorphism $E^2_{pq} \cong H_p(X, H_q(F))$) we used several times) with G' looking more than G accessible to further computational attacks in favourable cases. Nevertheless they may be important as theoretical background behind a genuine algorithm. [4], [8], [12], [14], [25] for example could give such theory suitable to the topic of this paper. In the more special areas of homology of iterated loop spaces or rational homotopy theory very extensive information can be received from [18] resp. [19].

4. Exercise: Show the extended calculability of the homology groups $H_i(\Lambda X)$, $i \geq 0$, with ΛX the free loop complex of the simply connected Kan-complex X having extended calculable homology groups $H_i(X)$, $i \geq o$. (Hence by the Kan-envelope procedure [10, p.36] the homology groups of the free loop space of a simply connected, finite simplicial complex are effectively calculable.)

References

[1]   H.J. Baues, Geometry of loop spaces and the cobar construction, Memoirs of the AMS, No. 230, 1980

[2]   P. Gabriel, M. Zisman, Calculus of Fractions and Homotopy Theory, Ergebnisse der Mathematik, Bd. 35, Springer-Verlag, Berlin 1967

[3]   A.D. Gavrilov, Effective computability of the rational homotopy type. Math. USSSR-Izv. 10, no. 6, 1976, 1239–1290 (Izv. Akad. Nauk SSSR, Ser. Mat. 40 (1976), no. 6, 1308–1331)

[4]   V.K.A.M. Gugenheim, On the perturbation theory of the homology of the loop space. J. Pure Appl. Algebra 25, no. 2, 1982, 197–203

[5]   K. Lamotke, Semisimpliziale algebraische Topologie. Die Grundlehren der math. Wissenschaften 147, Springer-Verlag, Berlin 1968

[6]   J.P. May, Simplicial Objects in Algebraic Topology, Van Nostrand Company 1967

[7]   J. McCleary, User's Guide to Spectral Sequences, Publish or Perish, Inc., Wilmington, Delaware (U.S.A.), 1985

[8]   J. McCleary, Cartan's cohomology theories and spectral sequences. Canadian Math. Soc. Conf. Proceedings, 1982. Current Trends in Algebraic Topology, Vol. 2, part 1, 499–506

[9]   R. Schön, A Five Lemma for Calculations in Homological Algebra, Memoirs of the AMS, this issue

[10]   R. Schön, An Algorithm for Calculating Homotopy Groups, Memoirs of the AMS, this issue

[11]   H. Schubert, Kategorien I, Springer-Verlag, Berlin 1970

[12]   V.A. Smirnov, Homology of fiber spaces, Russian Math. Surveys 35, no. 3, 1980, 294–298 (Uspehi Mat. Nauk 35 (1980), no. 3 (213), 227–230)

[13]   E.H. Spanier, Algebraic Topology, McGraw-Hill Book Company, New York 1966

[14]   J.C. Thomas, Eilenberg-Moore models for fibrations, TAMS 274, no. 1, 1982, 203–225

[15]   S. MacLane, Homology, Die Grundlehren der math. Wissenschaften 114, Springer 1967

[16]   D.G. Quillen, An Application of Simplicial Profinite Groups, Comm. Math. Helv. 44, 45–60, 1969

[17]   H.J. Baues, The double bar and cobar constructions, Compositio Mathematica 43, Fasc. 3, 1981, 331–341

[18]   F.R. Cohen, T.J. Lada, J.P. May, The homology of iterated loop spaces, Lecture Notes in Math. 533, Springer, Berlin 1976

[19]   D. Tanré, Homotopie Rationelle, Modèles de Chen, Quillen, Sullivan, Lect. Notes in Math. 1025, Springer, Berlin 1983

[20]   Wen-tsün Wu, de Rham-Sullivan Measure of Spaces and its Calculability. The Chern Symposium 1979, Berkely, Springer 1980

[21]   Wen-tsün Wu, Rational Homotopy Type, A Constructive Study via the Theory of the I^*-measure. Lecture Notes in Math. 1264, Springer, Berlin 1987

26 ROLF SCHÖN

[22] T. Kudo, S. Araki: Topology of H_n-Spaces and H-Squaring Operations. Mem. Fac. Sci. Kyushu Univ., Ser. A 10, 85–120 (1956)
[23] E.C. Zeeman, A Proof of the Comparison Theorem for Spectral Sequences. Proc. Camb. Phil. Soc. 53, 57–62 (1957)
[24] M.C. Tangora (editor), Computers in Geometry and Topology, Marcel Dekker, Inc. New York 1989
[25] L. Lambe, J. Stasheff, Application of perturbation theory to iterated fibrations, Manuscripta Math. 58, 363–376, 1989

Friedensstr. 2
6900 Heidelberg
Fed. Rep. of Germany

An Algorithm for Calculating Homotopy Groups

Abstract. This paper shows the effective calculability of each homotopy group $\pi_i(X), i \geq 1, X$ a simply connected complex of finite type, via the homology of iterated loop spaces. The key observation for the proof is the fact, that the classical Serre-class theory (Serre 1951) for fibrations can be generalized, so that extended calculability appears as a Serre-class property, and effective calculations of all needed spectral sequence terms and group extensions are possible. This has been shown in two previous papers. This one considers in detail the two special fibrations we need for our purposes: a path fibration, and another with fiber a $K(G, 1)$-complex. The only calculation results, which come in from outside, concern the homology of $K(\mathbb{Z}, 1)$- and $K(\mathbb{Z}/n\mathbb{Z}, 1)$-complexes. In order to get out their extended calculability, we supplement a special calculation method (Eilenberg-MacLane 1953). Another algorithm via the homology of Postnikov-complexes is pointed out, too. Some main problems (within algebraic topology) concerning improvements of efficiency are discussed. Good solutions are indispensable, if one wants to take a practical realisation into consideration.

I Recalling $K(G, 1)$-Complexes

Definitions, Notations:
The n-simplexes of $/G/ = K(G, 1)$ are by definition all $(n + 1) \times (n + 1)$-matrices A with coefficients $\underline{ij} \in G$, $0 \leq i, j \leq n$, such that $\underline{ij} + \underline{jk} = \underline{ik}$. Hence it is $\underline{ii} = 0$, $\underline{ij} = -\underline{ji}$, $/G/_o$ consists of one element (0), and an element (= matrix) $A \in /G/_n$, $n \geq 1$, can be characterized by its side-diagonal

$$(1) \quad (\underline{01}, \underline{12}, \dots, \underline{n-1, n}) .$$

The i-th boundary operator erases from A the row and the column with row-, resp. column-index i, and the differentials of the associated chain complex $C/G/$ in the representation (1) look like this:

$$d_1(x) = (0) - (0) = 0$$
$$d_2(x, y) = (y) - (x + y) + (x)$$
$$d_3(x, y, z) = (y, z) - (x + y, z) + (x, y + z) - (x, y)$$

in general:

$$d_n(\underline{1}, \underline{2}, \dots, \underline{n}) = \sum_{i=0}^{n} (-1)^i (\underline{1}, \underline{2}, \dots, \underline{i-1}, \underline{i+i+1}, \underline{i+2}, \dots, \underline{n}) ,$$

where the first and the last summand is understood to be $(\underline{2}, \underline{3}, \dots, \underline{n})$ resp. $(\underline{1}, \underline{2}, \dots, \underline{n-1})$.

1980 Mathematics Subject Classification (1985 Revision).
Primary 55 Q 05, 55 P 35, Secondary 55-04, 55 S 45

Classical Calculation Results:
(a) If $G = \mathbb{Z}$, then the identity $G \to \mathbb{Z}$, induces a homomorphism

$$h : C_1/G/ = \mathbb{Z}[G] \to Z \quad \text{with} \quad hd_2 = 0 \ ,$$

hence a homomorphism

$$(2) \quad \underline{h} : H_1/G/ \to \mathbb{Z} \ ,$$

which is easily proved to be an isomorphism. It is clearly $H_o/G/ \cong \mathbb{Z}$, and it is known, that $H_i/G/ = 0$, if $i \geq 2$.

(b) If $G = \mathbb{Z}/n\mathbb{Z}$, the group of integers $\{0, 1, 2, \ldots, n - 1\}$ with addition modulo a natural number $n \geq 2$, then the function

$$f : /G/_{2k+1} \to \mathbb{Z} \ ,$$

given by

$$f(\underline{1}, \bar{1}, \underline{2}, \bar{2}, \ldots, \underline{k}, \bar{k}, \underline{k+1}) = \underline{k+1}, \quad \text{if } \underline{i} + \bar{i} \geq n \quad \text{for all } i, \ 1 \leq i \leq n \ ,$$
$$= 0 \quad \text{else} \ ,$$

extends to a map of chain complexes $\underline{f} : C/G/ \to \underline{\mathbb{Z}}$ with $\underline{\mathbb{Z}}$ given by $\underline{\mathbb{Z}}_i = \mathbb{Z}$ for each $i \geq 0$. The differential $d : \mathbb{Z}_{i+1} \to \mathbb{Z}_i$, is given by multiplication with n, if i is odd, and $= 0$ else. The map $\underline{f}$ induces a homology isomorphism for all $i \geq 0$:

$$(3) \quad f_i = H_i(\underline{f}) : H_i/G/ \to H_i(\underline{\mathbb{Z}}) = \begin{cases} \mathbb{Z}/n\mathbb{Z} & \text{if } i \text{ is odd} \\ \mathbb{Z} & \text{for } i = 0 \\ 0 & \text{else} \end{cases}$$

(c) If G is a direct sum of copies from (a) and/or (b), then $H_i/G/$ can be calculated by an iterated application of the tensor-Tor formula because of the identity $/A \oplus B/ = /A/ \times /B/$ for groups A and B, which is immediate from the definition of $/G/$.

Concerning Literature:
The material above is extracted from [7] and [9]. For the calculation results a) and b) see [9, p.95 resp. p.97]. For the definitions see [7, p.485/486]. The following proposition concerning homotopy types of $K(G, 1)$-complexes is a variant of [7, p.492], and is true for arbitrary G, of one replaces $H_1(X)$ by $\pi_1(X)$. This generalization results from [7, p.492] by using the Milnor-homology-isomorphism $X \to S|X|$ [13, p.130].

Complexes Homotopy-Equivalent to $K(G, 1)$-Complexes
Let us assume, that X is a complex with the property $X = {}^1X$ (= first Eilenberg-subcomplex of X). Then each 1-simplex $x \in X_1$ is a cycle, and defines a homology class $[x] \in H_1(X)$. There is a canonical map

$$(4) \quad \text{can} : X \to /H_1(X)/ \ ,$$

which assigns to the single 0-complex of X_o the element $(0) \in /H_1(X)/_o$, and to an n-simplex $x \in X_n$, $n \geq 1$, the element $(\underline{01}, \underline{12}, \ldots, \underline{n-1, n}) \in /H_1(X)/_n$, where

$i - 1, i$, $1 \leq i \leq n$, is the homology class of this 1-face of x, which runs from the $(i - 1)$-th to the i-th corner; formally:

$$i - 1, i = i - 1, i(x) = [d_o d_1 \ldots \hat{d}_{i-1} \hat{d}_i \ldots d_n x]$$

(the first equality is a notation; a face operator with a roof must be left out). From now on: G finitely generated and abelian.

Proposition:

Let X be a Kan-complex with $X = {}^1 X$, and of the homotopy type of a $K(G, 1)$-complex. Then the canonical map

(4) $\operatorname{can} : X \to /H_1(X)/$,   described above,

a) is a homotopy-equivalence, and b) is an epimorphism.

Proof:

a) It is clear, that can, which is the canonical projection $X_1 \to /H_1(X)/_1$ in dimension 1, induces a homology epimorphism $H_1(X) \to H_1/H_1(X)/$, and hence (because between isomorphic groups, which are finitely generated, abelian, isomorphic to G) a homology isomorphism

$$H_1(X) \xrightarrow{\cong} H_1/H_1(X)/ ,$$

and hence (by the Hurewicz isomorphism) an isomorphism between the first homotopy groups. Because the higher homotopy groups are all trivial, we have therefore a weak homotopy equivalence between Kan-complexes, and hence a homotopy equivalence by a fundamental fact from the homotopy theory.

b) Let $\min X \to X$ be the inclusion of a minimal subcomplex $\min X$ of X, which is a homotopy equivalence. With a) the composition

$$\min X \to X \to /H_1(X)/$$

is a homotopy equivalence, and hence (because between minimal complexes) an isomorphism. Hence the second map is an epimorphism.

Complement:

Let us give a hint for an easy proof, that (4) is indeed a map between complexes: A skew-symmetric matrix A from $/G/$ is determined by its right upper triangular part, i.e. by a function $\underline{A} : \Delta[n]_1 \to G$, with $\Delta[n]_1$ the set of 1-simplexes of $\Delta[n]$, the standard n-simplex. Face and degeneracy operators of $/G/$ are then given by $\underline{A} \mapsto \underline{A} \cdot \delta^i$ resp. $\underline{A} \to \underline{A} \cdot \sigma^i$. $X \mapsto /G/$ above (with $G = H_1(X)$) is given in dimension n by sending $x \in X_n$ onto $\underline{A} : \Delta[n]_1 \xrightarrow[\underline{x}]{} X_1 \xrightarrow{q} H_1(X) = G$, with $\underline{x}$ induced by x and q the canonical projection. The standard switching rules for δ^i, σ^i, $\underline{x}$, see e.g. [13, p.3, p.11], imply immediately what we want.

II Calculating $K(G, 1)$-complexes

Proposition: (for calculating filtered $K(G, 1)$-complexes)
Let ${}_i X$, $i \geq 1$, be a preconstructible sequence [17, p.8(b)], which is at the same time a filtration of $X = /G/$ with union X. Then the following is true: $H_k(X)$ is extended calculable for each $k \geq 0$, if

(a) $G = \mathbb{Z}$ or

(b) $G = \mathbb{Z}/n\mathbb{Z}$,   or

(c) G is a direct sum of copies of (a) and/or (b).

Proof:
In view of the classical calculation results of section I, and the 2. Lemma of [17, p.3] it suffices to calculate in each case an epimorph integer. In all cases, where $H_k(X) = 0$, clearly $i = 1$ can be chosen as epimorph integer. In the cases, where $H_o(X) \cong \mathbb{Z}$, it is obvious, that the first integer i, such that ${}_i X \neq \emptyset$, is appropriate as epimorph integer. We now settle the remaining cases.

(a) An epimorph integer for $H_1(X)$.
INSTRUCTION: Run through the integers $i \geq 1$, and decide, whether $1 \in \mathbb{Z} = X_1$ is
    contained in ${}_i X_1$. Stop, if this is the case.
Because ${}_i X$ is an exhaustive filtration of X, the calculation stops after finitely many steps, say at the integer j. Then j is an epimorph integer: the composition

$$H_1({}_j X) \to H_1(X) \xrightarrow{\cong} \mathbb{Z} \,,$$

where the second map is the isomorphism $\underline{h}$ of **I** (2), maps the homology class [1] onto $1 \in \mathbb{Z}$, hence the first map is an epimorphism.

(b) An epimorph integer for $H_i(X)$, i odd.
INSTRUCTION: Run through the integers $j \geq 1$, and decide, whether the compo-
    sition $H_i({}_j X) \to H_i(X) \xrightarrow{\cong} \mathbb{Z}/n\mathbb{Z}$, where the second map is the
    isomorphism f_i of **I** (3), is an epimorphism. Stop, if this is the case.
Because $H_i(X)$ is the colimit of the groups $H_i({}_j X)$, $j \geq 1$, the calculation stops after finitely many steps yielding an integer j, which is then by instruction obviously epimorph.

Remark
It is possible to ignore the not so easily deducible classical result **I** b) by applying the fundamental theorem of [18] to the fibration

$$K(\mathbb{Z}, 1) \to K(\mathbb{Z}, 1) \to K(\mathbb{Z}/n\mathbb{Z}) \,,$$

where the first map is induced by multiplication with n. One has to construct an approximating sequence of maps, and may apply (a) above to have the extended calculability of the homology of the base and of the fiber. The extended calculability of the homology of the base follows then.

(c) is an immediate consequence of **I**(c) and the product algorithm [18, p.23, Remark 2].

Proposition: (for calculating homotopy types of $K(G, 1)$-complexes)
Let X be a Kan-complex having the homotopy type of a $K(G, 1)$-complex. If $H_1(X)$ is extended calculable, then each homology group $H_i(X)$, $i \geq 0$, is extended calculable.

Proof:
We may assume, that $X = {}^1X$, because otherwise we would apply the Five-Lemma Algorithm to the isomorphisms $H_i({}^1X) \xrightarrow{\cong} H_i(X)$, $i \geq 0$. Furthermore (by the 2. Remark of [17, p.6]) we may assume, that $H_1(X)$ has been calculated by minimal indexes a, b, i.e. we have the isomorphism ${}_aH_1(X) = \text{image}(H_1({}_aX) \to H_1({}_bX)) \xrightarrow{\cong} H_1(X)$, and for each $i \geq a$ the isomorphism

$$ {}_aH_1(X) \xrightarrow{\cong} {}_iH_1(X) \, , $$

where ${}_iH_1(X) = \text{image}(H_1({}_iX) \to H_1({}_jX))$, with j the minimal late isomorph index associated to i. With the proposition from **I** we have the upper horizontal isomorphism in the following diagram

$$
\begin{array}{ccc}
H_n(X) & \xrightarrow{\cong} & H_n/H_1(X)/ \\
\uparrow & & \uparrow \cong \\
H_n({}_iX) & \longrightarrow & H_n/{}_iH_1(X)/ \\
& & \uparrow \cong \\
& & H_n/{}_aH_1(X)/ \xrightarrow{\cong} H_n/TG/ \, ,
\end{array}
$$

with $G = {}_aH_1(X)$ and the obvious homomorphisms. Because $X \to /H_1(X)/$ is an epimorphism, the images ${}_i/G/ \subset /G/$ of ${}_iX$ under the composition

$$ {}_iX \to /{}_iH_1(X)/ \underset{\cong}{\leftarrow} /{}_aH_1(X)/ \to /G/ $$

form an exhaustive filtration of $/G/$. The subcomplexes ${}_i/G/$ form a preconstructible sequence, because for each element $x \in {}_iX_n$ each component

$$ \alpha(\underline{k-1}, k(x)) \in {}_iH_1(X) = \text{image}(H_1({}_iX) \xrightarrow{\alpha} H_1({}_jX)) $$

has a constructible image under the constructible composition

$$ {}_iH_1(X) \xleftarrow{\cong} {}_aH_1(X) \xrightarrow{\cong} G $$

(between finitely generated abelian groups). We have therefore a map ${}_iX \to {}_i/G/$ between preconstructible sequences with colimit the composition

$$ X \to H_1(X)/ \underset{\cong}{\leftarrow} /{}_iH_1(X) \underset{\cong}{\leftarrow} {}_aH_1(X) \underset{\cong}{\to} /G/ \, , $$

which induces a homology isomorphism for H_n. By the first proposition above (for calculating filtered $K(G, 1)$-complexes) $H_n/G/$ is extended calculable. Hence by the Five-Lemma Algorithm $H_n(X)$ is extended calculable.

III Calculating Homotopy Groups

Notation: ^2LX, second Eilenberg-subcomplex of LX, the loop-complex of X, as defined in [18, p.23].

Theorem:
Let X be a simply connected, preconstructible Kan-complex, such that each homology group $H_i(X)$, $i \geq 0$, is extended calculable. Then
   a) each homology group $H_i(^2LX)$, $i \geq 0$, is extended calculable,
   b) each homotopy group $\pi_n(X)$, $n \geq 1$, is effectively calculable.

Proof:
a) Let us consider the factorization of the path-fibration $PX \to X$ through its first Postnikov stage $1.PX$.

$$\begin{array}{ccc} ^2LX & \longrightarrow & PX \\ & & \searrow^u \\ \downarrow & & \quad 1.PX \longleftarrow 1:LX = F \\ & & \nearrow_v \\ & X & \end{array}$$

^2LX is the fiber of the fibration u with simply connected base $1.PX$, see [15, p.34, THEOREM 8.9 (iv)]. $1:LX = F$ is by definition the fiber of v, which has the homotopy type of an Eilenberg-MacLane complex $K(G,1)$ with $G = \pi_1(LX) \cong \pi_2(X) \cong H_2(X)$, see [15, p.35, COROLLARY 8.11].

Auxiliary statement: $H_1(F)$ is extended calculable. Proof: Because $H_1(F)$ is isomorphic to $H_2(X)$, $H_1(F)$ is clearly effectively calculable. Hence it suffices to calculate an epimorph integer. The first upper horizontal homomorphism in the following diagram is a monomorphism, because it is the first map of the factorization $\overset{\rightarrow}{\downarrow}$ of the obvious isomorphism $\downarrow_{\rightarrow}$

$$\begin{array}{ccccccc} \pi_2(1.PX, F) & \to & H_2(1.PX, F) & \overset{d}{\to} & H_1(F) \to H_1(1.PX) = 0 & \cdot \\ \cong\downarrow & & \downarrow p & & & \\ \pi_2(X, pt) & \underset{\cong}{\to} & H_2(X, pt) & & & \end{array}$$

It is also an epimorphism by the Four Lemma applied between the obvious exact sequences: regard, that we have the isomorphism $\pi_2(1.PX) \overset{\cong}{\to} H_2(1.PX)$, because $1.PX$ is simply connected, and the isomorphism $\pi_1(F) \underset{\cong}{\to} H_1(F)$, because $\pi_1(F) \cong H_2(X)$ is abelian. Hence the left vertical map p is an isomorphism (a spectral sequence argument can also be used for that). By the Five-Lemma Algorithm applied to p $H_2(1.PX, F)$ is extended calculable.

The appertaining epimorph integer can then be taken as epimorph integer for $H_1(F)$, because d is an epimorphism.

By the second proposition of **II** (for calculating homotopy types of $K(G,1)$-complexes) each homology group $H_i(F)$, $i \geq 0$, is extended calculable. By the algorithm for calculating the total space [18, p.22 Theorem (a)] each homology group of $1.PX$ is extended calculable. By the algorihm for calculating the fiber [18, p.22

Theorem (c)] each homology group of ^2LX is extended calculable, which finishes the proof.

b) is immediate by an inductive application of a), because $\pi_1(X) = 0$, and for $n \geq 2$ $\pi_n(X) \cong H_2(\underline{L}X)$ with $\underline{L}$ the $(n-2)$-fold iteration of ^2L.

1. Corollary: (for calculating homotopy groups of pairs)
Let (X, A) be a 2-connected pair of preconstructible Kan-complexes with A and X simply connected. If each homology group $H_i(A)$, $H_i(X)$, $i \geq 0$, is extended calculable, then each homotopy group $\pi_n(X, A)$, $n \geq 2$, is effectively calculable.

Proof:
Consider the cartesian square

$$
\begin{array}{ccccc}
LX & \longrightarrow & PX_{/A} & \longrightarrow & PX \\
 & & q \downarrow & & \downarrow p \\
 & & A & \longrightarrow & X
\end{array}
$$

with p the path fibration. By the algorithm for calculating the fiber LX of p each homology group $H_i(LX)$, $i \geq 0$, is extended calculable. Hence by the algorithm for calculating the total space of q, each homology group $H_i(PX_{/A})$, $i \geq 0$, is extended calculable. Because (X, A) is 2-connected, $PX_{/A}$ is simply connected. Hence by the theorem above each homotopy group $\pi_i(PX_{/A}) \cong \pi_{i+1}(X, A)$, $i \geq 1$, is effectively calculable.

2. Corollary: (for calculating iterated loop complexes)
Let X be a simply connected, preconstructible Kan-complex, such that each homology group $H_i(X)$ is extended calculable, then each homology group $H_i((\bar{L}X)_{pt})$, with $(\bar{L}X)_{pt}$ the base-point component of $\bar{L}X$ the n-fold loop complex of X, $n \geq 1$, is extended (hence effectively) calculable.

Proof:
An iterated application of Theorem a) above gives the extended calculability of $H_i(\underline{L}X)$, $i \geq 0$, with $\underline{L}$ the $(n-1)$-fold iteration of ^2L. Hence by Proposition a) of [18, p.24] (for calculating loop complexes) each homology group $H_i(L\underline{L}X)$, $i \geq 0$, is extended calculable. An application of the Five-Lemma Algorithm to the isomorphism $H_i((\bar{L}X)_{pt}) \to H_i(L\underline{L}X)$ induced by the homotopy equivalence and inclusion $(\bar{L}X)_{pt} \to L\underline{L}X$, implies the extended calculability of $H_i((\bar{L}X)_{pt})$, $i \geq 0$.

Remarks:
1. The existence of a sequence $_iX \to {}_{i+1}X$, $i \geq 1$, which is preconstructible (in principle) for every complex $_1X$ of finite type with colimit a Kan-complex is ensured by the Ex-construction of D.M. KAN [12], i.e. one chooses $_{i+1}X = \text{Ex}(_iX)$ inductively. The extended calculability of the colimit follows trivially (one chooses $i = 1$ as epimorph integer), because the inclusions $_iX \to {}_{i+1}X$ induce isomorphisms for all homology groups.

Closer to practicability is a different construction (namely the filling of a horn) in the following section **IV**, which can be considered as an effective and sparse version of the Kan-envelope-construction given by GABRIEL-ZISMAN [5, p.68].

2. With the 1. remark and case b) of the theorem above we have the effective calculability of the absolute homotopy groups $\pi_n(X)$, $n \geq 1$, for X simply connected and of finite type, a result, which is known by E.H. BROWN [2], who uses the Postnikov-system of X. Possibly his proof can be generalized to cover the effective calculability of homotopy groups of pairs by considering the Postnikov-system of the inclusion $A \to X$. A direct reduction to the single space case as above would require a finite model for $PX_{/A}$.

3. Our calculation possibilities for fiber spaces [18] allow one to calculate homology of $K(G,n)$-complexes as well as the homology of each Postnikov stage $n.X$ of a simply connected complex X of finite type, and we get immediately another algorithm for calculating homotopy groups, see section **V**. We defer to a different, separate note the effective calculability of the cohomology and of the k-invariants or Postnikov-complexes.

4. Existing literature chiefly concerns the possibility of homology calculations of iterated loop-suspension-spaces $\Omega^n \Sigma^n X$, see [4, p.490]. The 2. Corollary above (for calculating iterated loop spaces) generalizes the known effective calculability of the simple and the 2-fold loop-space, see [1].

5. In **IIIa**) we used tacitly, that the first **fibrewise** Postnikov-stage construction commutes with direct limits of sequences. Let us give the following categorical argument for the **absolute** case, which can easily be generalized.

By definition $(n.X)_k$ is the quotient of the set of all maps $u : \Delta[k] \to X$, defined by the equivalence relation $u \sim v : \Delta[k] \to X$, if u and v have the same restriction onto the n-skeleton $\Delta[k]^n \to X$. Equivalently $(n.X)_k$ is the push-out of the maps f, g in the following pull-back square

$$
\begin{array}{ccc}
E(X) & \xrightarrow{\;\;f\;\;} & hom(\Delta[k], X) = D(X) \\
{\scriptstyle g}\downarrow & \searrow\;(n.X)_k\;\hookleftarrow & \downarrow{\scriptstyle r} \\
D(X) = hom(\Delta[k], X) & \xrightarrow[r]{} & hom(\Delta[k]^n, X) = D^n(X)
\end{array}
$$

with r the obvious restricting function. Clearly D^n, D commute with colimits of sequences, hence so does the pull-back E, see e.g. [14, p.236/237]. Hence, so does $(n.)_k$, because itself a colimit.

6. Another, equivalent, more practical description of $(n.X)_k$ is given by defining k-simplexes $u, v \in X_k$ for $k > n$ as equivalent, if for each compositon $C : X_k \to X_n$ of face operators it is $Cu = Cv$, and it is $(n.X)_k = X_k$, if $k \leq n$.

Incidentally, it is an immediate consequence of [13, p.4 (1.4)], that it suffices to consider only those compositions $C = \underline{i}_1 \cdot \underline{i}_2 \ldots \underline{i}_s$ for which $i_k > i_{k+1}$, and the number of such compositions coincides with the number of strictly injective monotonous functions $[n] \to [k]$.

In our algorithm for calculating homotopy groups only $n = 1$ is needed; hence, from the viewpoint of complexity (in dependence of the dimension k), the construction of the first Postnikov-stage is unproblematic.

7. Let us formulate as exercises two simpler applications of the Five-Lemma Algorithm [17] and the fiber space calculations [18].

a) Let $F \subset G$ a pair of finitely generated, abelian groups (in type form). Show, that each homology group $H_i(G, F) = H_i(K(G, 1), K(F, 1))$ is effectively calculable.

b) Make precise and proof the proposition: each homology group $H_i(G)$ of a finitely generated, nilpotent group G is effectively calculable. (Hints: a group extension gives rise to a fibration, the precision concerns the form in which G is given.) The result is known even for polycyclic-by-finite groups, see [32].

IV The Horn-Filling Procedure

An n-dimensional subcomplex $S[n]$ of $\Delta[n+1]$, the standard $(n+1)$-simplex, generated by a subset $S \subset \Delta[n+1]_n$ is called a **horn** of dimension n, a **maximal horn**, if S consists of $n + 1$ non-degenerate simplexes. A map $s : S[n] \to X$ is called a **horn in** X or in X_n, and is called unfilled, if it has no extension (**filler**) onto $\Delta[n + 1]$. A **special horn** has a generating set S, which does not contain the first or the last face of $\Delta[n+1]$. X fulfills the **Kan-extension condition** up to dimension n, if every maximal, special horn up to dimension n has a filler, see [20] or [21]. The set of special horns in X_n directed by inclusion forms the **horn-graph** $Gr(X_n)$ of X_n. By the n-dimensional **homotopy-premodel** of X we mean the n pairs $(X_i, Gr(X_i))$, $1 \le i \le n$, together with a numbering (order) of all maximal horns occurring in these $Gr(X_i)$. We want to give an effective approximation to a Kan-envelope X_K of X; see [5, p.68] for this notion.

INSTRUCTION: (for filling a maximal, special, unfilled horn $s : S[n] \to X$):

Attach to X the standard $(n + 1)$-simplex $\Delta[n + 1]$ according to the pushout diagram

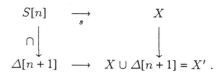

and: Modify $Gr(X_i)$ into $Gr(X_i')$ for $i \ge n$, by cancelling the horns in $Gr(X_i)$, which are no longer unfilled, and by adding the new arising ones. Those, which are new and maximal arrange after the old (ordered) maximal ones.

Explanations:

1. If X is a complex of finite type, it is an immediate consequence of [13, p.7, 3.9 Satz] that X is (numerically) representable by the following ingredients:

a) A finite set K of elements $\underline{1}, \underline{2}, \ldots, \underline{r}$, which are the non-degenerate simplexes of X.

b) A function dim : $K \to N$, which assigns to each simplex $\underline{i}$, $1 \le i \le r$, its dimension $i' = \dim(\underline{i}) \ge 0$.

c) For each $i, 1 \le i \le r$, a set of $i' + 1$ epimorphic, monotonous functions $[k, i]$, $0 \le k \le i'$, defined on $[i' - 1]$ by the equation

$$d_k(\underline{i}) = [k,i]^*\underline{a} \quad \text{for some} \quad \underline{a} \in K , \quad \text{i.e.} \ [k,i]^* : X_{a'} \longrightarrow X_{i'-1} ,$$

and $[k,i] : [i'-1] \to [a']$.

(The author has not considered more closely the problem how to represent such an X most conveniently. A representation generalizing the one of finite simplicial complexes, such that degenerate simplices appear not before they are needed should be desirable.)

2. Because $\Delta[n+1]$ can be constructed out of (a maximal horn) $S[n]$ by first attaching an n-simplex and then an $(n+1)$-simplex, it is clear, that the filling of a horn by the instruction above amounts numerically to the adding of two further elements $\underline{u}, \underline{v}$, $u = r+1$, $v = r+1$, to the set K of 1.a) above, and of appropriately defined monotonous functions according to 1.c).

3. It is a simple geometric observation, that (though the purpose of the instruction above is to fill a single horn) several other horns can be affected insofar as they lose their property of being unfilled or win the property of being maximal.

Obviously the instruction does not affect the graphs $Gr(X_i)$, $i \leq n-1$, and clearly the notation $i \geq n$ means, that the instruction is useful (in principle) up to an arbitrary high dimension i, which could be of interest in a practical or theoretical application.

Definition: (X a given complex of finity type)
By the n-dimensional **Kan-envelope procedure** we mean the following sequence:

$$Pre_k(X) = \{(_kX_i, Gr(_kX_i)/1 \leq i \leq n\} , \quad k \geq 1 ,$$

of homotopy-premodels:

$$Pre_1(X) = \{(X_i, Gr(X_i)/1 \leq i \leq n\} \ \text{is the premodel of} \ X , \quad \text{see above} .$$
$$Pre_{k+1}(X) \quad \text{is constructed out of} \ Pre_k(X) , \quad k \geq 1 ,$$

by filling according to the above INSTRUCTION the first (i.e. endowed with the smallest number) maximal horn of the horn-graphs of $Pre_k(X)$.

Proposition:
Let X be a complex of finite type, and $Pre_k(X)$, $k \geq 1$, its n-dimensional Kan-envelope procedure. Then
  a) The sequence $_kX$ is a preconstructible sequence in the sense of [17, p.8].
  b) The colimit complex fulfills the Kan-extension condition up to dimension n.
  c) For each $k \geq 1$ is the inclusion $_kX \to X' = colim(_iX)$ is a homotopy equivalence.
  d) Each homology group $H_k(X')$ is extended calculable with respect to these approximating complexes $_iX$.

Proof:
a) The constructibility of the graphs $Gr(X_i)$ and of the pushout (see the 2. Explanation above) immediately implies a).
b) Let s be a maximal, special horn in X' (of dimension $dim(s) \leq n$). Because s is of finite type, it must lie in some $_iX$, $i \geq 1$. We may assume, that s is unfilled in $_iX$ (otherwise the proof is ended), and hence that s belongs to a certain horn-graph

$Gr(_iX_k)$ with k appropriate, $1 \leq k \leq n$, endowed with a certain number. Because the Kan-envelope procedure fills all horns in the succession of their numbering (under keeping fixed all numbers of already numbered horns) this horn s will be filled in the colimit.

c) The cofibration and homotopy equivalence between the realizations $|S[n]| \rightarrow |\Delta[n+1]|$ implies a cofibration and homotopy equivalence between the realization $|_kX| \rightarrow |_{k+1}X|$, $k \geq 1$. The result follows from a standard fact in the homotopy theory of topological spaces, and the switching lemma for realizations with colimits.

d) Because of the isomorphism $H_k(_iX) \rightarrow H_k(X')$ for each $i \geq 1$, $k \geq o$, and $_1X = X$, $H_i(X')$ is effectively calculable, and an epimorph index is given by $i = 1$ (for all dimensions).

Remarks:

1. Let $X' \rightarrow X_K$ the inclusion of X' (from the Proposition c) above) into a Kan-envelope X_K (for all dimensions) of X by filling (inductively as in [5, p.68], not necessarily constructively) all and only the horns of dimension $> n$. Then X' up to dimension n is kept stationary, and is identical with X_K up to dimension n. This means, that general theorems require the Kan-condition only up to a certain dimension, when they involve only complexes up to this dimension.

2. Building up a horn-graph $Gr(X_n)$ is very time-consuming (if not improvable, one has to run through sets of subsets and to check up on matching conditions [15, p.2]; considering only the special horns is indicated by theory, otherwise $Gr(X_n)$ would be bigger yet). But regard, that this time-consumption is only in the beginning of the Kan-envelope procedure: as soon as a graph $Gr(_iX_n)$ is constructed, the adding of one further n-simplex to $_iX_n$ can create an unfilled horn only if added to a horn of $Gr(_iX_n)$, which has one element less, hence the check up on matching conditions is reduced drastically.

3. Naturally to consider only the unfilled horns has the effect, that the construction becomes stationary in dimensions $\leq n$, as soon as all horns of dimension $\leq n$ are filled. For example, if we start with a skeleton of the group-complex $K(G, 1)$, G finite, then $K(G, 1)$ up to a certain dimension is kept stationary throughout the procedure (in contrast to the Ex-construction of D.M. Kan [12]).

4. The efficiency problems of **VI** will convey the conviction, that the whole Kan-envelope procedure is superfluous as a not very practical method, and needed only theoretically for proofs, that certain complexes of finite type exist: if they are attached to a given one, the resulting complex fulfills certain homology conditions arising during the calculation process. Section **VI** will certainly make this remark more intelligible.

5. Concerning the 2. remark above, it is obvious, that the initial horn-graph $Gr(X_n)$ can be built up (at least partially) simultaneously with the input process, by, roughly speaking, sorting (or labelling) the input simplexes according to common faces.

V Calculating Homology of Postnikov-Complexes

Proposition   (for calculating homology of homotopy types of $K(G,n)$-complexes, G finitely generated, abelian):
Let X be a preconstructible Kan-complex having the homotopy type of a $K(G,n)$-complex. If $H_n(X)$ is extended calculable, then each homology group $H_i(X)$, $i \geq 0$, is extended (hence effectively) calculable. cf. [3], [8], [9].

Proof   (by induction on n):
The case $n = 1$ is the 2. Proposition of **II**. For the step from $n - 1$ to $n, n \geq 2$, we consider the path fibration

$$LX \to PX \to X ,$$

and may assume, that $H_n(X)$ is extended calculable. Hence by the Five-Lemma Algorithm applied to the isomorphisms

$$H_n(X, pt) \xleftarrow{\cong} H_n(PX, LX) \xrightarrow[d]{\cong} H_{n-1}(LX)$$

(use the diagram of the proof of the auxiliary statement of **IIIa**) with the obvious replacements), we have the extended calculability of $H_{n-1}(LX)$. Applying the induction hypothesis, each homology group $H_i(LX)$ is extended calculable. Hence, by the algorithm for calculating the base space [18, p.22 Theorem (b)], each homology group $H_i(X)$, $i \geq 0$, is extended calculable.

Theorem   (for calculating homology of Postnikov-complexes):
Let X be a simply connected, preconstructible Kan-complex, such that each homology group $H_i(X)$ is extended calculable. Then each homology group $H_i(n.X)$, $i \geq 0$, of each Postnikov-stage $n.X$, $n \geq 1$, is extended calculable.

Proof   (by induction on $n \geq 1$):
We use the following additional, auxiliary statement:
(n) The fiber $\underline{F}_n$ of the fibration $(n + 1).X \to n.X$ has the extended calculable
        homology group $H_{n+1}(\underline{F}_n)(\cong \pi_{n+1}(X))$.
For $n = 1$ $n.X$ is contractible, and has (trivially) extended calculable homology groups. Applying the Five-Lemma Algorithm to the isomorphisms

$$H_2(X) \xrightarrow[\cong]{} H_2(2.X) \xleftarrow[\cong]{} H_2(\underline{F}_1)$$

we have the statement (1). For the step from $n - 1$ to n, $n \geq 2$, we get first of all the extended calculability of the homology groups $H_i(\underline{F}_{n-1})$ by the proposition above, and hence by the algorithm for calculating the total space [18, p.22 Theorem (a)] the extended calculability of $H_i(n.X), i \geq 0$. In order to prove (n), we apply the algorithm for calulating the fiber [18, p.22 Theorem (c)] to the fibration

$$^{n+1}X \to X \to n.X , \quad [15, \text{p.32}] ,$$

^{n+1}X the $(n + 1)$-st Eilenberg subcomplex of X. The Five-Lemma Algorithm applied to the obvious isomorphism $H_{n+1}(\underline{F}_n) \xleftarrow[\cong]{} H_{n+1}(^{n+1}X)$ gives the desired result.

Remark

The proof of the Theorem above gives at the same time another algorithm for calculating homotopy groups: see the auxiliary statement (n) of the proof. It should be noted that it bypasses the hard classical problem of computing k-invariants.

VI Efficiency Problems

The purpose of this section is to point out some efficiency problems, which concern the complexity of the algorithm of **III** in dependence of n. They are caused by the enormous growing of the number of degenerate simplexes of X in high dimensions. As possible faces of individual horns they enter the horn-graphs $Gr(X_n)$, before few of them are turned into non-degenerate ones (by transitions from X to LX or to ^2X for example), and the rest of them can be neglected in calculating homology (by standard knowledge, see e.g. [16, p.193, Theorem 7.6]).

But the great flexibility of the Kan-envelope construction probably allows positive solutions: they consist in finding out how strong additional information arising during the calculation process can be utilized in order to consider and fill only those horns, which are really needed. Very roughly speaking, we have problems of this kind: how to kill homology classes by attaching cells (to certain allowed places) as efficiently as possible.

But first of all let us see, how the Five-Lemma Algorithm of [17] reduces to computations with indices in those cases, when the approximating sequences are exact themselves.

Lemma (generalizing the classical Five Lemma):
In the following diagram the rows are exact sequences of abelian groups.

$$
\begin{array}{ccccccccc}
''A & \rightarrow & 'A & \rightarrow & A & \rightarrow & A' & \rightarrow & A'' \\
\uparrow & & \uparrow & & \uparrow & & \uparrow & & \uparrow \\
''V & \rightarrow & 'V & \rightarrow & V & \rightarrow & V' & \rightarrow & V'' \\
\uparrow & & \uparrow & & \uparrow & & \uparrow & & \uparrow \\
''U & \rightarrow & 'U & \rightarrow & U & \rightarrow & U' & \rightarrow & U'' \\
\uparrow & & \uparrow & & \uparrow & & \uparrow & & \uparrow \\
''T & \rightarrow & 'T & \rightarrow & T & \rightarrow & T' & \rightarrow & T''
\end{array}
$$

a) If $'V \rightarrow 'A$, $U' \rightarrow A'$ are epimorphisms, and image$(U'' \rightarrow V'') \rightarrow A''$ is a monomorphism, then $V \rightarrow A$ is an epimorphism.

b) If image$('U \rightarrow 'V) \rightarrow 'A$, image$(T' \rightarrow U') \rightarrow A'$ are monomorphisms, and $''U \rightarrow ''A$ is an epimorphism, then image$(T \rightarrow V) \rightarrow A$ is a monomorphism.

c) Let the further exact sequence $''S \rightarrow 'S \rightarrow S \rightarrow S' \rightarrow S''$ be appended at the bottom of the diagram. Then by a) and b): if

$\mathrm{image}('T \to 'V) \to \mathrm{image}('U \to 'V) \to 'A$ and

$\mathrm{image}(S' \to U') \to \mathrm{image}(T' \to U') \to A'$ are isomorphisms,

$\mathrm{image}(S'' \to T'') \to A''$ a monomorphism, and

$''U \to ''A$ an epimorphism, then

$\mathrm{image}(T \to V) \to A$ is an isomorphism.

The proofs run by diagram chase as with the classical Five Lemma, and are omitted. Let us now assume, that the top exact sequence of the diagram above is approximated by (i.e. is the direct limit of) a sequence of exact sequences $('A'_i)$, $i \geq 0$.

$$
\begin{array}{ccccccccc}
''A & \to & 'A & \to & A & \to & A' & \to & A'' \\
\uparrow & & \uparrow & & \uparrow & & \uparrow & & \uparrow \\
''A_i & \to & 'A_i & \to & A_i & \to & A'_i & \to & A''_i
\end{array}
\qquad ('A'_i)
$$

Corollaries of the Lemma

a) Let $'a, a'$ be epimorph indices for $'A$ resp. A', and (a'', b'') early and late monomorph index for A'', such that $a' < a''$, then $\max('a, b'')$ is an epimorph index for A.

b) Let $('a, 'b), (a', b')$ be early and late monomorph index for $'A$ resp. A', and $''a$ epimorph index for $''A$, such that $''a < 'a, b' < 'a$, then $(a', 'b)$ are monomorph indices for A.

c) Let be

$('\underline{a}'\underline{b}), ('a, 'b)$ two pairs of isomorph indices for $'A$

$(\underline{a}', \underline{b}'), (a', b')$ two pairs of isomorph indices for A'

(a'', b'') monomorph indices for A''

$''a$ an epimorph index for $''A$ such that

$'\underline{a} < 'a, \quad a'' = \underline{a}' < a' = '\underline{a}, \quad b'' < a', \quad ''a < \max\{\underline{b}', b'\}$

Then $(a', \max\{'\underline{b}, 'b\})$ are isomorph indices for A.

The proofs use the fact, that if $i < j$ and i is epimorph (late monomorph, late isomorph) index, then so is j, see [17, p.2], and hence indices can be adjusted to have a diagram as needed in the lemma. The simple formulas of the OUTPUT-indices show, that the complexity, see e.g. [22, p.2], for getting them out, is determined by the one of providing INPUT-indices in the right size relationship. According to our definitions of monocal, epical, isocal, which yield arbitrarily big indices, such relationships can always be managed, and complexity considerations can be put into rigorous terms. We just remark, that our algorithms for calculating the total complex [17, p.8], the next base term [18, p.16], the next fiber term [18, p.18] have approximating exact sequences, whereas the following two not, which cause the efficiency problems mentioned in the beginning. They compute

1) The homology groups of F, a $K(G, 1)$-complex, see **II**.

2) E^2-terms of fibrations, see the 2. Proposition of [18, p.21].

In 1) and 2), the Five-Lemma Algorithm is applied to isomorphisms which are approximated by homomorphisms (which are no isomorphisms in general). Again we have a trivial specialization of the general Five-Lemma Algorithm, but now in a different direction:

By the implication (d) $\Rightarrow$ (c) of the 3. Lemma of [17. p.3] it suffices to calculate epimorph indexes. If $G \to H$ is an epimorphism, an epimorph index for G can be taken as epimorph index for H. The converse is obviously settled by the following instruction.

INSTRUCTION: Run through the indexes $i \geq 1$, and decide, whether the homomorphism $G(i) \to \text{image}(H('i) \to H(i'))$ is an epimorphism, where $i \mapsto ('i, i')$ is an isocalculation for H. Stop, if this is the case. (The appearing index is then epimorph for G.)

We have the efficiency problem of getting an epimorph index i as small as possible. By the following practical examples from IIIa), which are the first steps of the homotopy group algorithm, we can explain more closely how the general type of occurring efficiency problems looks like (we don't try to list them all, nor to formulate a general, comprehensive efficiency problem).

Example:

In IIIa) we have to calculate an epimorph index for $H_2(1.PX, LX)$. Because the filling of a horn to $_iX$ does not alter its homotopy type, an isocalculation for $H_2(X, pt)$ is given by $i \mapsto (i, i)$, and the instruction above specializes as follows:

INSTRUCTION: Run through the indexes $i \geq 1$, and decide, whether the homomorphism

$$(1) \quad H_2(1.P_iX, 1:L_iX) \to H_2(_iX, pt)$$

is an epimorphism. Stop, if this is the case.

The appertaining efficiency problem arises by considering the flexibility of the Kan-envelope construction, see **IV**, given by the freedom, in which succession the cells may be attached. (In fact any complex of finite type, which don't alter the homotopy type, may be attached, and the Kan-envelope construction can be considered as a proof, that there exists such a complex, which, if attached to $_1X$, the resulting complex makes (1) epimorphic. Leaving aside this generalization we have the following problem.)

1. Efficiency Problem:

Fill successively, as efficiently as possible, appropriate horns in $X = {}_1X, {}_2X, \ldots$ up to one gets a $_iX$ such that

$$H_2(1.P_iX, 1:L_iX) \to H_2(_iX, pt)$$

is an epimorphism.

(Clearly to be an epimorphism is equivalent to the vanishing of the cokernel, whence the phrase "killing homology classes" in the beginning of this section.)

Remark: Having calculated such an index i, which is at the same time an epimorph index for $H_1(1.LX)$, see IIa), the second step of the homotopy group algorithm, is to find an associated late isomorph index j for $H_1(1.LX)$, and the following efficiency problem arises.

2. Efficiency Problem:

Fill successively, as efficiently as possible, appropriate further horns in $_iX, _{i+1}X, \dots$ up to one gets a $_jX$ such that

$$H_1(1:L_iX) \rightarrow H_1(1:L_jX)$$

has as image a group isomorphic to $H_2(X)$.

Remark: Having calculated $H_1(F)$, $F = 1.LX$ a homotopy type of a $K(G,1)$-complex, the following problem arises in calculating all homology groups $H_i(F)$ $i \geq 0$, by the algorithms of **I, II**.

3. Efficiency Problem

Is it possible to calculate the homology groups $H_i(F)$, $i \geq 0$, such that the number $q(i)$ of horns needed for an isocalculation of $H_i(F)$ is polynomially bounded, i.e. $q(i) \leq p(i)$ for a certain polynomial $p(i)$?

Remark: The last problem we want to pose, concerns the 2. Proposition of [18, p.21], i.e. the calculation of E^2-terms, which might be feasible for appropriately chosen complexes $\underline{X}$.

4. Efficiency Problem:

For which finite complexes $\underline{X}$ can the $/n, p, 2/$-terms of the two fibrations $PX \rightarrow 1.PX \rightarrow X$ of section **III** be isocalculated by filling a polynomial bounded number $p(n)$ of horns?

Concluding Remarks:

1. The offered algorithm does without any work, e.g. [10], which tries to find out fine formulas for differentials of a fibration: these appear among the horizontal arrows in the diagram of the lemma, which do not enter the algorithm given by adjustment of indices as indicated above (after the corollaries). Nevertheless they can be computed simply as connecting homomorphisms between homology groups of finite (!) complexes: having computed the terms on the right by the groups on the left

$$(1) \quad \text{image}(_nE_p^r(_iX) \rightarrow {}_nE_p^r(_jX)) \xrightarrow{\cong} {}_nE_p^r(X) , \quad i \leq j ,$$

$$(2) \quad \text{image}(_{n-1}E_{p-r}^r(_iX) \rightarrow {}_{n-1}E_{p-r}^r(_jX)) \xrightarrow{\cong} {}_{n-1}E_{p-r}^r(X) , \quad i \leq j ,$$

the differential $d^r : {}_nE_p^r(X) \rightarrow {}_{n-1}E_{p-r}^r(X)$ is given as restriction of the connecting homomorphism

$$H_n(_jX^{p+r-1}, _jX^{p-1}) \rightarrow H_{n-1}(_jX^{p-1}, _jX^{p-r-1})$$

onto the left sides of (1) and (2), see section **III** of [17].

2. Introducing coefficients $\mathbb{Z}/n\mathbb{Z}$ into homology has several advantages: the parallel computing for different n, diagonalizing matrices only modulo n (hence avoiding the general case of integer matrices, which is very inefficient, [11], [23], [28]), and very probably the efficiency problems above can be solved by smaller complexes. A much more substantial and promising idea should be the discovering of small homology

models of the occurring chain complexes. Already their abstract existence should be an interesting theoretical result.

3. So far there exist no rigorous complexity results concerning homotopy group algorithms with simplicial sets as INPUT data. But too great hopes are not only damped by those with computational experiences in special cases [29], [30]. [24, p.210, Example] says, roughly speaking, that the size of homotopy groups can grow exponentially with the dimension for rather small complexes, see also [31]. Allowing Quillen's model (of a CW-complex X with one 0-cell and else cells only in dimensions 2 and 4) together with a natural number n as INPUT, it is shown in [6], that the problem of computing the n-th rational homotopy group $\pi_n(X) \otimes \mathbb{Q}$ is #P-hard. Notwithstanding that we use the mere "computability in principle" of homotopy groups and k-invariants [25] as computational prerequisites for classifying homotopy types effectively, see [26].

4. Open problem: Homotopy groups with coefficients, are they computable? One thinks of their importance in the homology decomposition of a space [27], constructed with help of k'-invariants. Are k'-invariants computable?

5. The computability of the homology groups of the free loop complex $\Lambda \underline{X}$ follows, as stated in exercise (4.) of [18], for $\underline{X}$ simply connected. It is unclear what happens for more general $\underline{X}$, and many open problems are connected with the homology with local coefficients. In particular, the extended calculability of the E^2-term of a fibration $F \to E \to B$ with non-simple coefficients $H_n(F)$ is unexplored.

References

[1]   H.J. Baues, The double bar and cobar constructions, Compositio Mathematica, 43, Fasc. 3, (1981), 331–341

[2]   E.H. Brown, Finite computability of Postnikov-complexes, Ann. of Math. (2) 65 (1957), 1–20

[3]   H. Cartan, Algèbra d'Eilenberg-MacLane et Homotopie, Séminaire 1954/55 d'Ecole Normale Supérieure, Paris, 1956

[4]   F.R. Cohen, T.J. Lada, J.P. May, The homology of iterated loop spaces, Lecture Notes in Math., vol. 533, Springer, Berlin, 1983

[5]   P. Gabriel, M. Zisman, Calculus of Fractions and Homotopy Theory, Ergebnisse der Mathematik, Bd. 35, Springer, Berlin, 1967

[6]   D.J. Anick, The Computation of Rational Homotopy groups is # P-hard, Lecture Notes in Pure and Applied Math. 114 (1989), 1–56, title: Computers in Geometry and Topology, editor: M.C. Tangora

[7]   S. Eilenberg, S. Mac Lane, Relations between homology and homotopy groups of spaces, Ann. of Math. (2) 46, (1945), 480–509

[8]   S. Eilenberg, S. Mac Lane, On the groups $H(\pi, n)$, (I) Ann. of Math. (2) 58, (1953), 55–106

[9]   S. Eilenberg, S. Mac Lane, On the groups $H(\pi, n)$, (II) Ann. of Math. (2) 60, (1954), 49–139

[10] T.V. Kadeisvili, The differentials of a spectral sequence of a twisted product, Sakharth. SSR Mecn. Akad., Moambe 82 (1976), no. 2, 285–288

[11] G. Havas, L. Sterling, Integer matrices and Abelian groups, Lecture Notes in Computer Science 72, Springer, Berlin, 1979, pp. 431–451, (Editor: E.W.Ng, Title: Symbolic and Algebraic Computation)

[12] D.M. Kan, On c.s.s. complexes, Amer. J. Math. 79 (1957), 449–476

[13] K. Lamotke, Semisimpliziale algebraische Topologie, Grundl. der math. Wissenschaften 114, Springer, Berlin, 1968

[14] S. Mac Lane, Kategorien, Hochschultext, Springer, Berlin, 1972

[15] J.P. May, Simplicial Objects in Algebraic Topology, Van Nostrand Company, New York, 1967

[16] A.T. Lundell, S. Weingram, The Topology of CW Complexes, Van Nostrand Company, New York 1969

[17] R. Schön, A Five Lemma for Calculations in Homological Algebra, Memoirs of the AMS, this issue

[18] R. Schön, Fibrations with Calculable Homology, Memoirs of the AMS, this issue

[19]   E.H. Spanier, Algebraic Topology, McGraw-Hill Book Comp., New York, 1966

[20]   S. Balcerzyk, On the Kan extension conditions, Bull. acad. Polon. Sci. Sér. Math. Astr. Phys. 24 (1976), 12, 1063–1066

[21]   Shih, Weishu, Sur la condition d'extension de Kan pour les complex semi-simpliciaux. C. R. Acad. Sci. Paris 244 (1957), 1131–1132

[22]   A.V. Aho, J.E. Hopcroft, J.D. Ullman, The Design and Analysis of Computer Algorithms, Addison-Wesley, Reading, Mass., 1974

[23]   R. Kamman, A. Bachem, Polynomial Algorithms for computing the Smith and Hermite Normal Forms of an integer Matrix, Siam J. Comput. 8, 1979, 499–507

[24]   K. Iriye, On the ranks of homotopy groups of a space, Publ. Res. Inst. Math. Sci. 23, 1987 no. 1, 209–213

[25]   R. Schön, The effective computability of k-invariants, Memoirs of the AMS, this issue

[26]   R. Schön, An effective classification of A_n^{n-1}-complexes, manuscript in preparation, Heidelberg, 1990

[27]   E.H. Brown and A.H. Copeland, "An homology analogue of Postnikov systems", Mich. Math. J. 6 (1959), 315–330

[28]   H. Lüneburg, On the computation of the Smith Normal Form, Lecture Notes in Computer Science, 296, (1987), 156–157, Trends in Computer Algebra (Symposium, Neuenahr 1987) Springer, Berlin 1987

[29]   D.C. Ravenel, Homotopy groups of spheres on a small computer, Lecture Notes in Pure and Appl. Math. 114 (1989), 259–287, Editor: M.C. Tangora, "Computers in Geometry and Topology"

[30]   M.C. Tangora, Computing the homology of the lambda algebra, Memoirs of the AMS 337 (1985)

[31]   Y. Felix, S. Halperin, J.C. Thomas, The homotopy Lie algebra for finite complexes, Publ. Math. IHES 56 (1982), 387–410

[32]   G. Baumslag, F.B. Cannonito, C.F. Miller, Computable Algebra and Group Embeddings, Journal of Algebra 69, (1981), 186–212

The Effective Computability of k-Invariants

Abstract. The note proves the effective computability of each k-invariant $k \in F(n.X)$, $n \geq 1$, of a simply connected, finitely generated complex X; $n.X$ its n-th Postnikov stage,

$$F(U) = H^{n+2}(U; TG) ,$$

TG the type of the homotopy group $\pi_{n+1}(X)$. More closely, an algorithm is provided, which constructs two finitely generated subcomplexes

$$n_1 \subset n_2 \subset n.X ,$$

and a cohomology class $\underline{k} \in F(n_1)$, such that the restriction $F(n.X) \longrightarrow F(n_1)$ induces an isomorphism

$$F(n.X) \overset{\cong}{\longrightarrow} \text{image}(F(n_2) \longrightarrow F(n_1)) ,$$

which maps the classical k-invariant $k \in F(n.X)$ onto $\underline{k} \in F(n_1)$.

"Simply finite complex" stands briefly for "1-connected, simplicial complex having finitely many non-degenerate simplices". The present note proves the computability of the k-invariants of such complexes. These lie in certain cohomology groups of the several Postnikov stages, the homology of which has been computed in [6, **V**]. An obstacle for dualizing the method of computation is the missing minimal property for descending chains of subgroups (of finitely generated abelian groups). We bypass this difficulty by computing cohomology out of homology via universal coefficient sequences. Section **I** proves the extended calculability of cohomology groups. Section **III** gives the main result by describing the algorithm. A crucial role plays the effective form of the classical fundamental class given in section **II** by the fundamental cocycle algorithm. The generalization to simple fibrations is pointed out in **IV**, after the basic lemma for our Postnikov-construction is proved in full generality. Section **V** shows, that equivalence between k-invariants is decidable.

Notations

Ab,	category of finitely generated, abelian groups
TG,	type of a group $G \in Ab$
$_iX, i \geq 1$,	a preconstructible sequence of complexes [5, **III**(b)], or chain complex with colimit X having finitely generated homology groups $H_i(X)$, $i \geq 0$.
$'C \oplus D$,	mapping cone of a chain map $u : C \to D$ with $('C \oplus D)_i = C_{i-1} \oplus D_i$ and boundary $\underline{d}(x, y) = (-dx, ux + dy)$
(Y, X),	(called 'pair of complexes') denotes the mapping cone $'CX \oplus CY$ if X and Y are complexes (or $('X \oplus Y$, if X and Y are chain complexes) with a $u : X \to Y$, which is obvious from the context.

1980 Mathematics Subject Classification (1985 Revision).
Primary 55 S 45, Secondary 55-04

I Calculating Cohomology Groups

The following lemma stands for the dual of the 1. Lemma of [5, p. 1]. Notation: $L = H^n(X; TG)$, $L(i) = H^n({}_iX; TG)$, $L(j, i) = \text{image}(L(j) \to L(i))$ for $j \geq i$.

1. Lemma

The (inverse directed) sequence of groups $L(i)$ has the following properties:
(1) There exists a monomorph integer $i \geq 1$, i.e. an integer $i \geq 1$, such that $L \to L(i)$ is a monomorphism.
(2) To each monomorph integer $i \geq 1$ there exists an integer $j \geq i$, such that $L \to L(j, i)$ is an isomorphism.

Proof

Both properties are proved with help of the following diagram having as rows the well known Ext-hom coefficient sequences for cohomology.

$$
\begin{array}{ccccccccc}
0 & \to & \text{Ext}(H_{k-1}(Y), TG) & \to & H^k(Y; TG) & \to & \text{hom}(H_k(Y), TG) & \to & 0 \\
& & \downarrow & & \downarrow & & \downarrow & & \\
0 & \to & \text{Ext}(H_{k-1}({}_jY)TG) & \to & H^k({}_jY; TG) & \to & \text{hom}(H_k({}_jY), TG) & \to & 0
\end{array}
$$

(1) Choose $k = n$, $Y = X$, ${}_iY = {}_iX$ with $j = i$ an integer, which is split epimorph for $H_{n-1}(X)$ and epimorph for $H_n(X)$. By the classical Five-Lemma $H^n(X; TG) = L \to L(i) = H^n({}_iX; TG)$ is a monomorphism.
(2) Let i be a fixed monomorph integer. Choose now $Y = (X, \underline{X})$ with $\underline{X} = {}_iX$, $k = n + 1$, and ${}_jY = ({}_jX, \underline{X})$ with $j \geq i$ appropriately chosen by (1), such that the map $\underline{j}$ in the following diagram is a monomorphism.

$$
\begin{array}{ccccc}
H^n(X; TG) & \overset{i'}{\to} & H^n\ (\underline{X}; TG) & \overset{d'}{\to} & H^{n+1}(X, \underline{X}; TG) \\
\downarrow & {}^{i}\nearrow & & \searrow d & \downarrow \underline{j} \\
H^n({}_jX; TG) & & & & H^{n+1}({}_jX, \underline{X}; TG)
\end{array}
$$

Then $\text{kernel}(d') = \text{kernel}(d)$ implies $\text{image}(i') = \text{image}(i)$ or equivalently the isomorphism

$$
H^n(X; TG) = L \overset{\cong}{\to} L(j, i) = \text{image}(H^n({}_jX; TG) \to H^n(\underline{X}; TG)) .
$$

Remark

Let ${}'AB$ denote the category of (inverse directed) sequences (L) of finitely generated, abelian groups $L(i)$, $i \geq 1$, with limit L, having the properties (1) and (2) of the 1. Lemma above. Observation: partial material of section **I** of [5] has a dualization onto ${}'AB$. We give here only the following useful definitions and facts and refrain from a categorial preparation.

Definitions   An object (L) in ${}'AB$ is called **precalculable**, if there exists an algorithm, which assigns to an integer $i \geq 1$, the types $TL(i + 1)$, $TL(i)$ and a homomorphism

$TL(i+1) \to TL(i)$, such the object TL is isomorphic to L in $'AB$. (L) is called epi- (mono-, iso-) calculable, briefly **epical** (**monocal, isocal**), if there exists an algorithm, which assigns to an integer $i \geq 1$ a pair of integers j, k with $i \leq j \leq k$, such that one of the following equivalent conditions (a), (b) is fulfilled.

(a) $L \to L(k)$ induces an epi- (mono-, iso-) morphism $L \to L(k, j)$.

(b) For each $l \geq k$ $L \to L(l)$ induces an epi- (mono-, iso-) morphism $L \to L(l, j)$.

j and k are called the **early resp. late epimorph** (monomorph, isomorph) index associated to i.

Obvious facts: (c) If (L) is monocal, then it is monocal with the same early and late monomorph index. (d) If (L) is isocal, then it is monocal and epical. (e) If (L) is isocal, then with k a late isomorph index $L \to L(k)$ is a split monomorphism. k is then called **split monomorph** index.

(L) is called **extended calculable**, if the limit L of (L), an early isomorph index j, and an associated late isomorph index k is calculable. The triple (k, j, TL) is called **extended value** of (L). k, l are called **isomorph indices** for (L).

2. Lemma

An object (L) in $'AB$ is monocal, if and only if a monomorph integer is calculable.

3. Lemma

For a precalculable object (L) in $'AB$, the following properties are equivalent:

(f) (L) is isocal

(g) (L) is monocal and epical

(h) (L) is extended calculable

(i) An extended value of (L) is calculable.

We omit the simple proofs of the 2. and 3. Lemma, which are dual to the ones of [5, p.3] with exception of one point: as for (h) $\Rightarrow$ (f) we cannot conclude dual to [5, p.3], that the monomorphism $L \to L(a'(i), 'a(i))$ between isomorphic, finitely generated, abelian groups is an isomorphism. Here the strengthened extended calculability from above enters. Minimal indexes are defined by the following instructions. (k, j, TL) denotes an extended value of (L).

1. INSTRUCTION:

Run through the (finite) decreasing sequence of integers j', $1 \leq j' < j$, and decide, whether $L(k, j) \to L(k, j')$ is isomorphic. Stop, if this is not the case.

2. INSTRUCTION:

Run through the (finite) decreasing sequence of integers $'k, a \leq 'k < k$, and decide, whether $L(k, j) \to L('k, a)$ is isomorphic. Stop, if this is not the case.

Explanations:

If the 1.Instruction terminates at an index $j' > 1$, we choose $a = j' + 1$ as minimal early isomorph index, and else, i.e. at $j' = 1$, we choose $a = 1$ or $a = 2$ according to $L(k, j) \to L(k, 1)$ is isomorphic or not. This a enters the 2.Instruction, which may terminate at k'. If $k' > a$, we choose $b = k' + 1$ as minimal late isomorph index, and else, i.e. for $k' = a$, we choose $b = a$ or $b = a + 1$ according to $L(k, j) \to L(a, a)$ is isomorphic or not.

Proposition   (for calculating cohomology groups)

If the homology groups $H_{n-1}(X)$, $H_n(X)$, $H_{n+1}(X)$ are extended calculable, then the cohomology group $H^n(X;TG)$ is extended calculable.

Proof

By assumption we can calculate an integer i, which is split epimorph for $H_{n-1}(X)$ and epimorph for $H_n(X)$. Then $L \to L(i)$ is a monomorphism, see the proof of (1) of the 1. Lemma. Let be $\underline{X} = {}_iX$. By the Five-Lemma algorithm applied to the obvious part of the exact sequence of the pair $(X, \underline{X})$, the homology groups $H_{n+1}(X, \underline{X})$ and $H_n(X, \underline{X})$ are extended calculable; hence we can calculate an integer j, which is epimorph for the former and split epimorph for the latter group. By the proof of (2) of the 1. Lemma $L \to L(j, i)$ is an isomorphism.

Theorem   (for calculating cohomology of Postnikov complexes)

Let X be a simply connected Kan-complex, such that each homology group $H_i(X)$, $i \geq 0$ is extended calculable. Then each cohomology group $H^i(n.X; TG)$, $i \geq 0$, of each Postnikov stage $n.X$, $n \geq 1$ is extended calculable.

Proof

By the theorem of [6, V] (for calculating homology of Postnikov complexes) each homology group $H_i(n.X)$ is extended calculable. Hence by the proposition above, so is each cohomology group $H^i(n.X; TG)$.

II The Fundamental Cocycle Algorithm

Let us bring to mind the numerical procedure, which calculates the homology group $H_n(C; Z) = H_n(C)$, C a free chain complex of finite type, with help of the following diagram. TG denotes the type of $G = H_n(C)$.

$$
\begin{array}{ccc}
C_{n+1} \xrightarrow{\;V\;} C_{n+1} & & \\
\downarrow d' & \searrow D' & \\
C_n \xrightarrow{S} C_n \xrightarrow{r} Z_n \xrightarrow{Q} Z_n \xrightarrow{p} TG \\
\downarrow d \quad \downarrow D & & \\
C_{n-1} \xrightarrow{U} C_{n-1} & &
\end{array}
$$

Explanation (assuming throughout, that C_n are direct sums of copies $\mathbb{Z}$)

By the matrices S and U the boundary d is diagonalized into D. kernel$(D) = Z_n$ is a direct sum of summands of C_n. r is the projection. By the matrices Q and V the composition rSd' is diagonalized into D'. p is the projection.

Obvious facts and definition:

The composition $F = pQrS : C_n \to TG$ is a cocycle $F \in Z^n(C; G)$ and induces an isomorphism $\underline{F} : H_n(C) \to TG$. F is called the numerical **fundamental cocycle** of $H^n(C; H_n(C))$. More generally for any group $G \in Ab$, an element $K \in Z^n(C; TG)$ is called a (**numerical**) **cocycle** of $H^n(C; G)$.

Lemma (extension property for cocycles, C, K, G as above)

Let $C' \to C$ be a chain map with $C'_{n-1} \to C_{n-1}$ an inclusion of a direct summand, and J', K' homologous cocycles in $Z^n(C'; TG)$, with K' the restriction of K. Then there exists an effectively calculable cocycle $J \in Z^n(C; TG)$, which is homologous to K and extends J'.

Proof (each C_i is assumed to be given as a direct sum of copies $\mathbb{Z}$)

Consider the following diagram

$$
\begin{array}{ccc}
\hom(C_{n-1}, TG) & \xrightarrow{\;d\;} & Z^n(C; TG) \\
\downarrow & & \downarrow \\
\hom(C'_{n-1}, TG) & \xrightarrow{\;d'\;} & Z^n(C'; TG)
\end{array}
$$

By assumption the equation $d'(X') = J' - K'$, which is numerically a system of linear diophant equations, has a solution $X' : C'_{n-1} \to TG$, which is effectively calculable (by classical general theory) and has an obvious extension $X : C_{n-1} \to TG$. Choose $J = K + d(X)$; its restriction is $K' + (J' - K') = J'$.

Definition ($_iC$, $i \geq 1$ a sequence of chain complexes, C the colimit)

An **effective fundamental cocycle** F of $H^n(C; G)$, $G = H_n(C)$ is a cocycle $_iF \in Z^n(_iC; TG)$ with i appropriate, such that its cohomology class $[_iF]$ extends uniquely to a classical fundamental class $[F] \in H^n(C; TG)$, meaning, that the canonical map $H^n(C; TG) \to \hom(H_n(C), TG)$ maps $[F]$ onto an isomorphism $H_n(C) \to TG$.

Proposition (for calculating effective fundamental cocycles)

Let $_iC$, $i \geq 1$, be a preconstructible sequence of chain complexes with colimit C and extended calculable homology groups $H_i(C)$, $i \geq 0$. Then an effective fundamental cocycle F of $H^n(C; H_n(C))$ is calculable.

Proof (Notations: $G = H_n(C)$, $G(i) = H_n(_iC)$, $G(i, j) = \text{image}(G(i) \to G(j))$)

By the Proposition of **I** (for calculating cohomology groups) $H^n(C; TG)$ is extended calculable. Let us choose (effectively calculated) indices $a \leq b \leq c \leq d$, such that a, b are isomorph indices for $H_n(C)$, b, c are isomorph indices for $H^n(C; TG)$, and c, d again isomorph indices for $H_n(C)$. Let $'F$ be a numerical fundamental cocycle of $H^n(_cC; H_n(_cC))$ as defined above, and let F' be its image in $Z^n(_bC; TG(c, d))$ under the following composition with the first map induced by $Ts : TG(c) \to TG(c, d)$, the evaluation of $G(c) \to G(c, d)$.

$$
'F \in Z^n(_cC; TG(c)) \to Z^n(_cC; TG(c, d)) \to Z^n(_bC; TG(c, d)) \ni F'.
$$

We have the following diagram with the obvious homomorphisms.

$$
\begin{array}{ccc}
[F] \in H^n(C; TG) & \xrightarrow{\;v'\;} & \hom(H_n(C), TG) \\
= \uparrow & & \uparrow = \\
H^n(C; TG(c, d)) & \longrightarrow & \hom(H_n(C), TG(c, d)) \\
\downarrow & & \downarrow \cong \\
['F] \in H^n(_cC; TG(c)) \xrightarrow{\;u\;} H^n(_cC; TG(c, d)) & \xrightarrow{\;v\;} & \hom(G(a, b), TG(c, d)) \\
\downarrow & & \uparrow \\
[F'] \in H^n(_bC; TG(c, d)) & \longrightarrow & \hom(H_n(_bC), TG(c, d))
\end{array}
$$

Hence it suffices to show, that vu maps $['F]$ onto an isomorphism $w : G(a,b) \to TG(c,d)$. It is easily shown, that w is this composition:

$$G(a,b) \underset{t}{\to} H_n({}_bC) \underset{r}{\to} H_n({}_cC) \quad \underset{\underline{F}}{\overset{\cong}{\to}} \quad TG(c) \quad \underset{Ts}{\to} \quad TG(c,d)$$

$$s \searrow \qquad \qquad \cong \nearrow f$$

$$G(c,d)$$

where Ts is the calculated s, i.e. it is defined by the triangle with f determined by the calculating algorithm of $G(c,d)$. Because $srt : G(a,b) \to G(c,d)$ is an isomorphism, the result follows: $[F]$, the uniquely determined preimage of $[F']$, is mapped by v' onto an isomorphism $H_n(C) \to TG$.

Definition (the Fundamental-Cocycle algorithm)

As in the proposition let ${}_iC, i \geq 1$, be a preconstructible sequence of chain complexes with colimit C and extended calculable homology groups $H_i(C), i \geq 0$.

A fundamental cocycle algorithm is an algorithm, which assigns to integers $i \geq 1$ cocycles $F/i/ \in Z^n({}_iC; TG)$ with the following properties:
(1) $F/i/$ is the restriction of $F/i+1/ \in Z^n({}_{i+1}C; TG) \to Z^n({}_iC; TG)$.
(2) If $j \geq i$ are isomorph indices for $H^n(C; TG), G = H_n(C)$, then the isomorphism

$$H(j,i) = \text{image}(H^n({}_jC; TG) \to H^n({}_iC; TG)) \overset{\cong}{\leftarrow} H^n(C; TG)$$

maps the cohomology class $[F/i/] \in H^n({}_iC; TG)$ onto a fundamental class $[F] \in H^n(C; TG)$ i.e. onto a class $[F]$ inducing an isomorphism $H_n(C) \to TG$.

A Fundamental-Cocycle algorithm exists if ${}_iC_{n-1} \to {}_{i+1}C_{n-1}, i \geq 1$, **are inclusions of direct summands**: by the proposition above we have for the special isomorph indices $(j,i) = (c,b)$ a cocycle $F/b/ \in Z^n({}_bC; TG)$ with the property as stated in (2). Clearly for indices $j, 1 \leq j < b$, the cocycles $F/j/$ may be defined by restrictions. Let us assume, that $F/j/$ is calculated for an index $j \geq b$. By the implication (h) $\Rightarrow$ (f) of the 3. Lemma we can calculate isomorph indices (k',k) for $H^n(C; TG)$ with $k' \geq k \geq j + 1$. With $l = \max\{k', c\}$ we have the obvious isomorphism $H(l,k) \overset{\cong}{\to} H(c,b)$. By the Lemma of section II $F/b/$ can be extended to a cocycle $U \in Z^n({}_lC; TG)$. We choose as $F/k/, j + 1 \leq k \leq l$, the restrictions of U onto $Z^n({}_kC; TG)$.

III Postnikov Complexes and k-Invariants

Let ${}_if : {}_iY \to {}_iZ$ be a preconstructible sequence of maps [5, III(b) p.8], such that ${}_iY \to {}_{i+1}Y$, ${}_iZ \to {}_{i+1}Z$ are inclusions. By the previous section II we have the fundamental cocycle algorithm, which provides numerical cocycles $F/i/ \in Z^n({}_iC; TG)$ with $G = H_n(C)$, C now the mapping cone ${}'CY \oplus CZ$ associated to the colimit map $f : Y \to Z$ and ${}_iC$ the corresponding ones to ${}_if : {}_iY \to {}_iZ$. The (abstractly existing) cocycle $F \in Z^n(C; TG)$, defined by its restrictions $F/i/$ onto ${}_iC$, has as induced map an isomorphism $\underline{F} : H_n(C) \to TG$, hence is a fundamental cocycle in the classical sense. The restrictions of F onto ${}'CY$ and CZ define a cochain $J \in C^{n-1}(Y; TG)$ resp. a cocycle $K \in Z^n(Z; TG)$, and hence the maps j and k in the following diagram.

$$j(y) = \bar{y}(J) \qquad \begin{array}{ccc} Y_l & \xrightarrow{j} & C^{n-1}(\Delta[l]; TG) \xleftarrow{\bar{y}} C^{n-1}(Y; TG) \ni J \\ f \downarrow & & \downarrow d \end{array}$$

$$k(z) = \bar{z}(K) \qquad \begin{array}{ccc} Z_l & \xrightarrow{k} & Z^n(\Delta[l]; TG) \xleftarrow{\bar{z}} Z^n(Z; TG) \ni K \end{array}$$

As it is well known and easily proved, the square is a commutative square of (simplicial) complexes with dimension index l. It is an immediate consequence of the definition of F, that the square restricted onto $_if$ is the one associated to the cocycle $F/i/$.

Notations

$E(k, d)$ the fiber product of k with d.

$E(_ik, {_id})$ the fiber product of $_ik$ = (restriction of k onto $_iZ$)
              with $_id$ = (restriction of d onto image($_ij$)),
              where $_ij$ = (restriction of j onto $_iY$).

Lemma

If $Y \to E(k, d)$ is surjective, then $E(_ik, {_id})$ from above is an exhaustive filtration of $E(k, d)$, and hence $_iY \to E(_ik, {_id})$ is a preconstructible sequence of maps, which approximate $Y \to E(k, d)$.

Proof

Because Y is the colimit of the $_iY$, the images $E(_ik, {_id})$ of the maps $_iY \to Y \to E(k, d)$ form an exhaustive filtration of $E(k, d')$, d' the restriction of d onto image(j). Because $Y \to E(k, d') \subset E(k, d)$ is surjective, it is $E(k, d') = E(k, d)$.

The following equivalent description of $E(k, d')$ is obviously constructive for finite, constructible complexes Y, Z, and gives the rest.

The preconstructible Postnikov stage $E(k, d')$ (see square above): image(j) is isomorphic to $Y_l / \underline{J}$ with $\underline{J}$ an equivalence relation on Y_l given by "$y \sim y'$, if for each $(n-1)$-face Δ' of $\Delta[l]$ it is $J(y) = J(y')$ for the restrictions $\underline{y}, \underline{y}'$ of y resp. y' onto Δ'." $E(k, d')$ is therefore isomorphic to the subcomplex $E(F)$ of $Z \times (Y/\underline{J})$, defined by

$$E(F)_l = \left\{ \begin{array}{l} \text{set of these } (z, [y]) \in Z \times (Y/\underline{J}) \text{ such that for each } n\text{-face} \Delta'' \\ \text{of } \Delta[l] \text{ the } K\text{-value } K(\underline{z}) \text{ is equal to the alternating sum} \\ \sum(-1)^r J(\underline{y}^r), \ o \le r \le n, \text{ with } \underline{z} \text{ the restriction of } z \text{ onto } \Delta'', \\ \text{and } \underline{y}^r \text{ the restriction of } y \text{ onto the } r\text{-th } (n-1)\text{-face of } \Delta''. \end{array} \right\}$$

Definitions and Notations

A **canonical Postnikov tower** up to stage n, denoted by $P(n)$ is a tower of fibrations

$$P(n) \quad n.P \xrightarrow{'n'} n - 1.P \longrightarrow \ldots \xrightarrow{'2'} 1.P \xrightarrow{'1'} 0.P = * \,,$$

such that the fibration $'n'$ is induced by a map $n.k : n-1.P \to K(Tn.G, n+1)$, called the n-th **k-invariant**, with $Tn.G$ the type of a group $n.G \in Ab$, from the fibration $E(Tn.G, n) \xrightarrow{d} K(Tn.G, n+1)$. The complexes

$$E(Tn.G, n)_l = C^n(\Delta[l], Tn.G) , \qquad K(Tn.G, n+1)_l = Z^{n+1}(\Delta[l], n.TG)$$

are understood to be normalized (see [3, p.100/101]). $P(n)$ is called a **tower for a complex** X, if the standard (functorial) tower $X(n)$ of X up to stage n is homotopy equivalent to $P(n)$.

$X(n)$ $n.X \longrightarrow n-1.X \longrightarrow \ldots \longrightarrow 1.X \longrightarrow 0.X = *$.

That is, we have homotopy equivalences $m.f : m.X \to m.P$, $o \leq m \leq n$, which present a tower map $f(n) : X(n) \to P(n)$. (Incidentally, the latter condition implies that $Tn.G$ must then be isomorphic to the n-th homotopy group $\pi_n(X)$.) For $\underline{X}$ finite, X denotes a Kan envelope of $\underline{X}$.

Theorem (constructing Postnikov complexes, calculating k-invariants)
Let $\underline{X}$ be a simply finite complex, then for each $n \geq o$ a canonical Postnikov tower $P(n)$ for X is preconstructible, i.e. each stage $m.P$, $o \leq m \leq n$, has a preconstructible, exhaustive filtration $i.m.P.$, $i \geq 1$, such that $i.m.P. \to i.m-1.P$, $1 \leq m \leq n$, is induced by $i.m.k.$, the restriction of the m-th k-invariant $m.k : m-1.P \to K(T\pi_n(X), m+1)$ onto $i.m-1.P$, and $i.m.k$ is effectively calculable as cohomology class of $H^{m+1}(i.m-1.P; T\pi_n(X))$.

Proof (by induction on $n \geq o$)
$n = o$ is trivial. Let $_iX, i \geq 1$, with $_1X = \underline{X}$ denote the effective Kan-envelope procedure of [6, p.36] with colimit X. Then the maps $_iX \to _{i+1}X$ are inclusions and so are the induced maps $n._iX \to n._{i+1}X$. Let $P(n-1)$ for X be preconstructible, $n \geq 1$. We use the following statement as additional hypothesis:

$ADD(n-1)$: A homotopy equivalence $n-1.f : n-1.X \to n-1.P$
is preconstructible by approximating maps
$n-1._if : n-1._iX \to i.n-1.P$.

By the theorem of [6, p.38] (for calculating homology of Postnikov complexes) each homology group $H_i(n-1.X)$, $H_i(n.X)$, $i \geq o$, is extended calculable, hence by the Five-Lemma algorithm so is each homology group $H_i(n-1.P)$, $H_i(n-1.P, n.X)$, $i \geq o$, where the pair $(n-1.P, n.X) = C$ stands for the mapping cone constructing of the composition $n.X \to n-1.X \to n-1.P$ (as defined in the lines before section I). By the Proposition of **I** (for calculating cohomology groups) the cohomology group $H^{n+1}(C; TG)$ with $G = H_{n+1}(C)$ is extended calculable (here $ADD(n-1)$ enters). By section **II** and the following second additional hypothesis $\underline{ADD}(n-1)$ we have therefore the fundamental cocycle algorithm assigning to $i \geq 1$ the cocycle

$F/i/ \in Z^n(_iC; TG)$ with $_iC = (i.n-1.P, n._iX)$.

$\underline{ADD}(n-1)$: $n-1.P$ is approximated by inclusions
$i.n-1.P \to i+1.n-1.P$, $i \geq 1$, of subcomplexes

Let us consider the following square, where the maps j, k, and the cocycle K are explained in the beginning of this section **III**.

$$
\begin{array}{ccc}
n.X & \xrightarrow{\;j\;} & E(TG, n) \\
\downarrow & & \\
n-1.X \quad n.P & & \Big\downarrow d \\
\downarrow & & \\
k(x) = \bar{x}(K) \qquad n-1.P & \xrightarrow{\;k\;} & K(TG, n+1) \xleftarrow{\;\bar{z}\;} Z^{n+1}(n-1.P; TG) \ni K
\end{array}
$$

By classical theory we know, that the canonical map $n.f : n.X \to n.P$ into the fiber product $n.P = E(k, d)$ is a homotopy equivalence, and hence an epimorphism, because $n.P$ is minimal. We define now $i.n.P = E(_ik, _id)$, with $_ik$ the restriction onto $i.n - 1.P$ and $_id$ the restriction onto $j(n._iX)$. By the lemma above the canonical maps $n._iX \rightarrow i.n.P$ present a preconstructible sequence, which approximate $n.f$: $n.X \rightarrow n.P$. Hence $ADD(n)$ is proved. By definition of $E(_ik, _id)$ $i.n.P \rightarrow i.n - 1.P$ is induced by $_ik$ being the restriction of the n-th k-invariant $k = n.k$ (which is the first part of the theorem). It is a general property of a fiber product, that inclusions $j(n._iX) \rightarrow j(n._{i+1}X)$ and $i.n - 1.P \rightarrow i + 1.n - 1.P$ induce an inclusion $i.n.P \rightarrow i + 1.n.P$. Hence $\underline{ADD}(n)$ is proved.

By the representation theorem for cohomology we know, that the cocycle K (see above) defines the cohomology class appertaining to k. Hence its restriction

$$K/i/ \in Z^{n+1}(i.n - 1.P; TG) ,$$

which is effectively given as the restriction of $F/i/$, represents the cohomology class we are looking for. It is computed by the following instruction.

INSTRUCTION: Calculate the image $[k/i/]$ of $[F/i/]$ under the homomorphism (between cohomology of finite complexes)

$$[F/i/] \in H^{n+1}(i.n - 1.P, n.X; TG) \longrightarrow H^{n+1}(i.n - 1.P; TG) \ni [k/i/] .$$

This proves the second part of the theorem.

Application (for X a Kan-complex with $H_i(X)$, $i \geq 0$, extended calculable)
Let X be simply connected and have a finite, known number of non-vanishing homotopy groups. It is effectively decidable, whether X has the homotopy type of an abelian group complex.

Proof (with the simple observation, that in the theorem above, '$\underline{X}$ simply finite' may be replaced by 'X simply connected with extended calculable homology groups')
Let $b(n)$ be an early isomorph index for the cohomology group $H^{n+1}(n - 1.P; TG)$; $G = \pi_n(X)$. By above we can calculate each k-invariant

$$k/b(n)/ \in H^{n+1}(b(n).n - 1.P; TG) \overset{i}{\leftarrow} H^{n+1}(n - 1.P; TG) \ni k ,$$

which is the image of the classical k-invariant k under the monomorphism i. Hence we must only see, whether the finitely many classes $k/b(n)/$ for which $\pi_n(X)$ is different from zero, vanish.

The result then follows by a classical theorem (e.g. [2, p.240]): If this is the case, then X has the homotopy type of an abelian group complex; otherwise not.

Remark:
The application has no significance, if we would assume a simply finite $\underline{X}$ instead of X: It is a known result of J.R. Hubbuck [8], that a simply finite complex has the homotopy of an abelian H complex, if and only if it is contractible. (Hence non-trivial examples can be received out of fiber space constructions with help of the calculation possibilites of [9].)

IV k-Invariants of Simple Fibrations

The previous section can be generalized to simple fibrations, yielding the calculability of the (co-)homology groups and of the k-invariants of each Postnikov-stage. A specialized version of the following lemma was already needed in the proof of the **Theorem** of that section (see the corollary below). It is the crucial lemma, on which our special kind of Postnikov-construction is based. All complexes are assumed to be connected.

1. Lemma $(E \xrightarrow{P} B$ a Kan-fibration with fiber $F, n \geq 1)$
Let F be $(n-1)$-connected. The following square associated to a fundamental cocycle of

$$Z^{n+1}(\text{Cone}(p); TG) \ , \ \text{Cone}(p) = {}'CE \oplus CB \ , \ TG \text{ type of } G = H_{n+1}(\text{Cone}(p)) \ ,$$

as in section **III**,

$$
\begin{array}{ccc}
E & \xrightarrow{\ j\ } & C^n(\Delta[-]; TG) = \underline{C} \\
(p_k^j d) \quad p \downarrow & & \downarrow d \\
B & \xrightarrow{\ k\ } & Z^{n+1}(\Delta[-]; TG) = \underline{Z} \ ,
\end{array}
$$

has this property:
The restriction of j between the fibers $F \to K = Z^n(\Delta[-]; TG) = K(TG, n)$ induces an isomorphism

(1) $H_n(F)_{/\pi_1(B)} \xrightarrow{\ \cong\ } H_n(K(TG, n)) \cong TG$.

Proof
With help of the canonical fundamental cocycle of d it is easy to see, that the map pair j, k induces an isomorphism

(2) $H_{n+1}(\text{Cone}(p)) \xrightarrow{\ \cong\ } H_{n+1}(\text{Cone}(d))$.

The decomposition of p into an inclusion $E \xrightarrow{\ i\ } B'$, and a homotopy equivalence $B' \to B$, a fibration, gives the following chain of isomorphisms

$$H_{n+1}(\text{Cone}(p) \cong H_{n+1}(\text{Cone}(i)) \cong H_{n+1}(B', E) \cong H_o(B, H_{n+1}(cF, F))$$
$$\cong H_{n+1}(cF, F)_{/\pi_1(B)} \cong H_n(F)_{/\pi_1(B)} \ .$$

The third one is by the spectral sequence of the fibration $(B', E) \to B$ with fiber (cF, F), cF the (contractible) fiber of $B' \to B$. The fourth one is the well known formula for homology with local coefficients in dimension 0 yielding the group of orbits of $H_{n+1}(cF, F)$. The other ones are obvious.

The corresponding chain of isomorphisms with p replaced by d yield $H_{n+1}(\text{Cone}(d)) \cong H_n(K)$, $\underline{Z}$ being simply connected. Hence (2) is translated into (1).

Corollary   ($A \in Ab, n \geq 1$, B a Kan-complex)

In the lemma above let F have the homotopy type of a $K(A, n)$-complex, and $E \to B$ be simple, i.e. $\pi_1(B)$ operates trivially on $H_n(F)$. Then in the square $(p_k^j d)$ of the lemma p is fiber homotopy equivalent to the fibration induced by k from d. The fiber homotopy equivalence is given by the canonical map

$$f : E \longrightarrow E(k, d) \ (= \text{fiber product of } B \text{ and } \underline{C} \text{ over } \underline{Z}) \ .$$

Proof

By the lemma f induces an isomorphism $H_n(\underline{f})$, $\underline{f}$ the restriction of f between the fibers, hence isomorphisms $\pi_i(\underline{f})$ and $\pi_i(f)$, $i \geq 1$. Therefore f is a homotopy equivalence and hence a fiber homotopy equivalence by the simplicial version of [11, (6.21) p.119].

2. Lemma   ($p : E \to B$ a Kan-fibration between Kan-complexes)

Let us be given a square $p_k^j d$ as in the 1. Lemma, such that the canonical map $f : E \to E(k, d)$, see above, is a fiber homotopy equivalence (no further assumptions). Then f is epimorphic.

Proof

Let $'E \to B$ be a minimal subfibration of p, such that the inclusion i in the following diagram is a fiber homotopy equivalence.

$$'E \overset{i}{\longrightarrow} E \overset{f}{\longrightarrow} E(k, d)$$
$$\searrow \quad \downarrow \quad \nearrow$$
$$B$$

Because $d : \underline{C} \to \underline{Z}$ is minimal, $E(k, d) \to B$ is minimal. Hence fi is a fiber homotopy equivalence between minimal fibrations and must be an isomorphism. Hence f is epimorphic.

Remarks

a) The corollary above implies that the map k is indeed a representative of the classical characteristic class of the fibration (defined as transgression of the fundamentalclass of F): A classical representative k' has the same implication, namely a fiber homotopy equivalence $f' : E \to E(k', d)$. Hence the fibrations $E(k, d) \to B \leftarrow E(k', d)$ are fiber homotopy equivalent, and the homotopy classes of k and k' are equivalent, i.e. the one can be transformed into the other by a coefficient automorphism $TG \to TG$. See section VIII.4. of [2] for the classical theory.

b) Another implication of the corollary is the following: if the coefficients $H_n(F)$ are simple, then so are all $H_i(F)$, $i \geq 0$. This is true, because the local coefficients $H_i(F)$ on B are induced by the local coefficients $H_i(K)$ on $\underline{Z}$ being simple, because $\underline{Z}$ is simply connected.

c) We define in this note an arbitrary fibration $F \to E \to B$ as simple, if $\pi_1(F)$ is abelian, and each Postnikov-fibration $n.E \to n - 1.E$ has fiber F_n with simple coefficients $H_i(F_n)$, $i \geq 0$. With help of b) and the isomorphism $H_n(F_n) \cong \pi_n(F)$, it is easy to show, that $E \to B$ is then a simple map (as defined in [12, p.447], the definitions are not quite uniform in the literature).

d) The following theorem generalizes the one of section **V** of [6] from simply connected complexes to simple fibrations. As a side result we get the succeeding computational theorem for homotopy groups of simple complexes X, i.e. of complexes X, such that $X \to pt$ (= one point) is a simple fibration as defined in c) or – what is in this special case equivalent –, which are simple in the classical sense.

1. Theorem   (calculating (co-)homology of Postnikov-stages, $G \in Ab$)
Let $F \to E \to B$ be a simple fibration with extended calculable homology groups $H_i(F)$, $H_i(B)$, $i \geq 0$. Then each (co-)homology group $H_i(n.E, TG)$ resp. $H^i(n.E, TG)$, $i \geq 0$, of each (fiberwise) Postnikov-stage $n.E, n \geq 0$, is extended calculable.

Proof (by induction on $n \geq 0$)
Because F is connected, its 0-stage $0.F$, the fiber of $0.E \to B$, is contractible. Hence $0.E \to B$ is a homotopy equivalence, and $H_i(0.E)$, $i \geq 0$, is extended calculable.

Let each homology group $H_i(n-1.E)$, $i \geq 0$, be extended calculable. In the following vertical map between fibrations the coefficients $H_n(F.n) \xrightarrow{\cong} H_n(K.n)$ are simple; they are calculable by the general computational theorem for fibrations applied to (1)

$$
\begin{array}{ccccc}
(1) & F_n & \longrightarrow & E & \longrightarrow & n-1.E \\
 & \downarrow & & \downarrow & & \downarrow \\
(2) & K.n & \longrightarrow & n.E & \longrightarrow & n-1.E
\end{array}
$$

$K.n$ has the homotopy type of a $K(\pi_n(F), n)$-complex. Hence by [6, p.38 Prop.] each homology group $H_i(K.n)$ is extended calculable. By the general computational theorem for fibrations applied to (2) each homology group $H_i(n.E)$, $i \geq 0$, is extended calculable. Hence so is each cohomology group $H^i(n.E; TG)$, $i \geq 0$, by section **I**, TG the type of the finitely generated abelian group G.

2. Theorem   (calculating homotopy groups of simple complexes)
Let $\underline{X}$ be a finitely generated, simple complex. Then each homotopy group $\pi_i(\underline{X})$, $i \geq 1$, is effectively calculable.

Proof
We have the preconstructible Postnikov-stages $n.X$ with X a preconstructible Kan-envelope of $\underline{X}$, see section **IV** of [6]. By above $H_n(F.n)$, which is isomorphic to $\pi_n(X)$, is extended calculable. Hence $\pi_n(\underline{X}) = \pi_n(X)$ is effectively calculable.

Remark
We defer to a separate note the calculability of the homotopy groups of finitely generated, nilpotent complexes. Here the full 1. Lemma with a non-trivial action of $\pi_1(B)$ comes in.

Definitions and Notations
A canonical Postnikov tower (over B) up to stage n, denoted by $P(n)$, is a tower of fibrations

$$n.P \xrightarrow{'n'} n - 1.P \longrightarrow \ldots \longrightarrow 1.P \longrightarrow 0.P = B \ ,$$

such that the fibration $'i'$, $1 \leq i \leq n$, is induced by a map

$$i.k : i - 1.P \to K(Ti.G, i + 1) \ ,$$

called the i-th k-invariant, with $Ti.G$ the type of a group $i.G \in Ab$, from the fibration

$$E(Ti.G) = C^n(\Delta[\ ], Ti.G) \xrightarrow{d} Z^{n+1}(\Delta[\ ], Ti.G) = K(Ti.G, i + 1)$$

between complexes understood to be normalized.

$P(n)$ is called a tower for the fibration $E \to B$, if the standard, functorial tower

$$n.E \longrightarrow n - 1.E \longrightarrow \ldots \longrightarrow 1.E \longrightarrow 0.E \ ,$$

is homotopy equivalent to the tower $P(n)$ above, meaning, that we have homotopy equivalences $i.f : i.E. \to i.P$, $0 \leq i \leq n$, with $0.f : 0.E \to B$ the projection, which present a tower map.

3. Theorem (constructing Postnikov-towers, calculating k-invariants)

Let $F \to E \to B$ be a simple, preconstructible fibration, E, B Kan-complexes with extended calculable homology groups $H_i(E), H_i(B)$, $i \geq 0$. Then for each $n \geq 0$ an appertaining canonical Postnikov-tower $P(n)$ is preconstructible, i.e. each stage $i.P$, $0 \leq i \leq n$, has a preconstructible exhaustive filtration $j.i.P$, $j \geq 1$, such that $j.i.P \to j.i - 1.P$ is induced by

$$j.i.k : \ j.i - 1.P \longrightarrow K(Ti.G, i + 1) \ ,$$

the restriction of the i-th k-invariant $i.k : \ i - 1.P \to K(Ti.G, i + 1)$, which is effectively calculable in the sense, that there exists calculable indices $a \leq b$, such that $a.i - 1.P \to i - 1.P$ induces an isomorphism

$$H^{i+1}(i - 1.P; Ti.G) \xrightarrow{\cong} \text{image}(H^{i+1}(b.i - 1.P; Ti.G) \to H^{i+1}(a.i - 1.P; Ti.G))$$

mapping k onto an effectively calculable cohomology class

$$\underline{k} \in H^{i+1}(a.i - 1.P; Ti.G) \ , \quad i.G. = \pi_i(F) \ .$$

Proof

With the 1. and 2. Lemma above the corresponding proof of section **III** has an almost word for word generalization. The 1. Theorem above replaces the Theorem of section **I**.

Remark

It is a common requirement for computations, that a computational result ought to be independent of the algorithm applied. This cannot be achieved here. The k-invariant depends on the fundamental cocycle algorithm chosen, and different algorithms yield equivalent k-invariants, i.e. the one can be transformed into the other by a coefficient automorphism, see Remark a), p.55. It is therefore of special interest that this equivalence is decidable, proved in the next section.

V Equivalence of k-Invariants is Decidable

1. Lemma

Let F be a direct sum of k copies $\mathbb{Z}$, $m \geq 2$ an integer. An automorphism $A :$ $F/mF \rightarrow F/mF$ has an automorphism $\underline{A} : F \rightarrow F$ as extension, if and only if $\det(A) = \pm 1 (\mathrm{mod}\, m)$.

Proof (cf. [7, p.36/37], $A = \underline{A}(\mathrm{mod}\, m)$ means $a_{ij} = \underline{a}_{ij}(\mathrm{mod}\, m)$)
If $\underline{A} : F \rightarrow F$ is an automorphism, it is $\det(\underline{A}) = \pm 1$. Because $A = \underline{A}(\mathrm{mod}\, m)$, it is $\det(\underline{A}) = \pm 1 (\mathrm{mod}\, m)$.

The converse: let D be the Smith diagonal form of $A = UDV$, U, V integer matrices with $\det(UV) = 1$. We construct below a homomorphism $C : F \rightarrow F$ with $C = D(\mathrm{mod}\, m)$ and $\det(C) = \pm 1$; then the proof is ended by choosing $\underline{A} = UCV$: it is $\det(\underline{A}) = \det(U)\det(C)\det(V) = \det(UV)\det(C) = \pm 1$, hence $\underline{A}$ is an automorphism, and furthermore we have the $(\mathrm{mod}\, m)$-equality $\underline{A} = UCV = UDV(\mathrm{mod}\, m) = A(\mathrm{mod}\, m)$. Hence $\underline{A}$ extends A.

Construction of C: Let be $D = \mathrm{diag}(\underline{1}, \underline{2}, \ldots, \underline{k})$, $i \in \mathbb{Z}$, and $\det(A) = \det(D) = \underline{1} \cdot l = \pm 1 + dm$, with $l = \underline{2} \cdot \underline{3} \cdot \cdot \underline{k}$, and d appropriate.
Since $(l, m) = 1$, there exist integers s, t, such that

$$d + ls + m^{k-1}t = 0 .$$

Let T be the matrix of order $k \times k$ with entries $t_{11} = s$, $t_{k,1} = t$, and $= 0$ else. Choose $C = D + m(S + T)$, S the side diagonal matrix $s_{i,i+1} = 1$, $1 \leq i \leq k - 1$, $= 0$ else. Then obviously $C = D(\mathrm{mod}\, m)$, and by a simple calculation

$$\det(C) = (\underline{1} + ms)l + m^k t = \pm 1 + m(d + ls + m^{k-1}t) = \pm 1 .$$

2. Lemma

Let $'F, F'$ be direct sums of j resp. k copies $\mathbb{Z}$. $F = {'F} \oplus F'$. $A : F/mF \rightarrow F/mF$ an automorphism. Then an automorphism $\underline{A} : F \rightarrow F$, which extends A and the inclusion $'F \rightarrow F$, exists, if and only if the restriction of A onto $'F/m'F$ is the inclusion $'F/m'F \rightarrow F/mF$, and the composition $U : F'/mF' \rightarrow F/mF \xrightarrow{A} F/mF \rightarrow$ F'/mF' (inclusion follwed by A and the projection) has $\det(U) = \pm 1(\mathrm{mod}\, m)$.

Proof
Let as first of all remark, that any homomorphism $\underline{A} : F \rightarrow F$, which is the inclusion on $'F$, has the form

$$\underline{A} = \begin{pmatrix} E & \underline{V} \\ 0 & \underline{U} \end{pmatrix} \quad \begin{matrix} E, \text{ the } j \times j \text{ unit diagonal matrix, } V \text{ a } j \times k \text{ matrix ,} \\ 0, \text{ the } k \times j \text{ zero matrix, } \quad \underline{U} \text{ a } k \times k \text{ matrix ,} \end{matrix}$$

and it is $\det(\underline{A}) = \det(\underline{U})$.

If $\underline{A}$ is an automorphism, then $\det(\underline{U}) = \pm 1$, and hence by the 1. Lemma $\det(U) = \pm 1(\mathrm{mod}\, m)$, and clearly, if $\underline{A}$ extends $A : F/mF \rightarrow F/mF$, then A restricted onto $'F/m'F$ is the inclusion.

The converse: if $\det(U) = \pm 1(\mathrm{mod}\, m)$, then there exists by the 1. Lemma an extension $\underline{U} : F' \rightarrow F'$ of U (with $\det(\underline{U}) = \pm 1$). We now choose as extension

$\underline{V}: F' \to F$ an arbitrary extension of the composition $V: F'/mF' \to F/mF \xrightarrow{A} F/mF \to {'F}/m'F$ (e.g. we may choose $\underline{V} = V$ as matrices). Then obviously the matrix $\underline{A} = \begin{pmatrix} E & \underline{V} \\ 0 & \underline{U} \end{pmatrix}$ is an automorphism extending A and the inclusion ${'F} \to F$.

Remark

We use here the notion 'extension' in a double sense: In the one case a map is extended from a direct summand onto the whole group (leaving the image space fixed). In the other case, we consider any map as extension of an induced one between corresponding quotient groups.

3. Lemma

Let ${'F}, F'$ be direct sums of j resp. k copies $\mathbb{Z}$, $F = {'F} \oplus F'$, $I : {'F} \to F$ a monomorphism, and $A : F/mF \to F/mF$ an automorphism. It is effectively decidable, whether there exists an automorphism $\underline{A} : F \to F$, which extends I and A (simultaneously).

Proof

Because any automorphism $\underline{A} : F \to F$ preserves direct sum decompositions, the following instruction is obvious:

INSTRUCTION: Decide, whether $I({'F})$ is a direct summand of F, i.e. (equivalently) whether the quotient $F/I({'F})$ is free.

If the decision is answered positively, we continue, and note, that then the Smith diagonal form $D : {'F} \to F$ of $I = \underline{U}D\underline{V}$ is the inclusion of the direct summand ${'F}$.

$$
\begin{array}{ccc}
{'F} & \xrightarrow{\ \ D\ \ } & F \\[4pt]
\underline{V} \big\uparrow & & \big\downarrow \underline{U} \\[4pt]
{'F} & \xrightarrow{\ \ I\ \ } & F
\end{array}
$$

$\underline{A}$ is an extension automorphism of I and A, if and only if $\underline{B} = U^{-1}\underline{A}\tilde{V}^{-1}$ is an extension automorphism of D and $B = U^{-1}AV^{-1}$ with

$$
\tilde{V} = \begin{pmatrix} V & 0 \\ 0 & E \end{pmatrix} , \quad E \text{ the } k \times k \text{ unit matrix, hence } \tilde{V}D = \underline{V}D ,
$$

where $U, V : F/mF \to F/mF$ is induced by $\underline{U}$ resp. $\tilde{V}$.

Now we are in the situation of the 2. Lemma. It decides, whether D and B can be extended, and the proof is ended.

4. Lemma

It is effectively decidable, whether the integer matrix equations

a) $\begin{aligned} XD &= R \\ X &= S \,(\mathrm{mod}\, m) \end{aligned}$ D a diagonal matrix $\mathrm{diag}(\underline{1}, 2, \ldots \underline{j}, 0 \ldots)$

b) $\begin{aligned} YA &= B \\ Y &= C \,(\mathrm{mod}\, m) \end{aligned}$

have solutions X, Y, which are automorphisms.

Proof

  a) Because XD is the matrix X with the first j columns multiplicated by the integers $i, 1 \le i \le j$, the following instruction is obvious.

INSTRUCTION: Decide, whether the first j columns of R are divisible by
$1, 2, \ldots, j$ respectively, and the other ones vanish.

If this is the case, a solution of a) corresponds to an automorphism $X : F \to F$ (= sum of $\dim(X)$ copies of $\mathbb{Z}$), which extends $S : F/mF \to F/mF$ and the homomorphism $T : {}'F \to F$, given by $T({}'i) = i'/i$, $1 \le i \le j$, where ${}'F$ is the direct sum of j copies of $\mathbb{Z}$, ${}'i$ the i-th canonical base vector, and i' the i-th column of R.

Because restrictions of automorphisms are monomorphisms:

INSTRUCTION: Decide, whether the homomorphism $T : {}'F \to F$ (just described) is a monomorphism.

If this is the case, we are in the situation of the 3. Lemma, and the proof is ended.

  b) With the equivalent equation system $YU^{-1}UAV = BV$, $YU^{-1} = CU^{-1} (\mathrm{mod}\, m)$, $UAV = D$ the Smith diagonal form of A, we have a reduction to a) by choosing $R = BV$, $S = CU^{-1}$, $X = YU^{-1}$.

Corollary

Let F be a direct sum of copies $\mathbb{Z}$, $\underline{C} : F/mF \to F/mF$ an automorphism, and ${}'i, i' \in F$, $1 \le i \le j$, j pairs of elements. It is effectively decidable, whether there exists an automorphism $C : F \to F$, which extends $\underline{C}$, and maps ${}'i$ onto i'.

Proof

This is a reformulation of b) of the 4. Lemma, choosing as columns for A and B the j elements ${}'i$ resp. i'. $\dim(F)$ gives the number of rows.

Proposition

Let TG be the type of a finitely generated abelian group G, and $j \ge 2$, $1 \le j \le s$, integers. ${}'i, i' \in TG$, $1 \le i \le r$, ${}''j, j'' \in TG/jTG$, $1 \le j \le s$, are pairs of elements. It is effectively decidable, whether there exists an automorphism $A : TG \to TG$, which maps ${}'i$ onto i' and ${}''j$ onto j''.

Proof

The splitting of $TG = F \oplus \underline{F}$ into the free part F and finite part $\underline{F}$, induces the splitting $TG/mTG = F/mF \oplus \underline{F}/m\underline{F}$, and a splitting of elements $x \in TG$ or $x \in TG/mTG$ into components, $x = (f(x), \underline{f}(x))$. Each automorphism $A : F \oplus \underline{F} \to F \oplus \underline{F}$ has the form $A(f, \underline{f}) = (U(f), V(\underline{f}) + W(f))$, with U, V automorphisms of F resp. $\underline{F}$, and W a homomorphism $F \to \underline{F}$. Because $Aut(\underline{F})$ is finite and enumerable, see [7], and likewise (trivially) $\hom(F, \underline{F})$, the following instruction is well-stated.

INSTRUCTION: Run through the pairs of homomorphisms $V \in Aut(\underline{F})$, $W \in \hom(F, \underline{F})$. Stop, if $V(\underline{f}({}'i)) + W(f({}'i)) = \underline{f}(i')$ for $1 \le i \le r$, and $\underline{V}(\underline{f}({}''j)) + \underline{W}(f({}''j)) = \underline{f}(j'')$ for $1 \le j \le s$.

$\underline{V} \in Aut(\underline{F}/j\underline{F})$, $\underline{W} \in \hom(F/jF, \underline{F}/j\underline{F})$ denote the homomorphisms induced by V resp. W.

If the instruction yields no appropriate homomorphisms V, W, the decision of the proposition is answered negatively. Otherwise we may take the resulting V, W for defining the component $\underline{f}(A(x, y))$ as $V(y) + W(x)$, and the algorithm continues to decide, whether an appropriate $C \in Aut(F)$ exists. Any $C \in Aut(F)$ induces an automorphism $\underline{C} : F/mF \to F/mF$, m the least common multiple of the numbers $\underline{j}$, $1 \leq j \leq s$. Hence a C we are looking for must have an induced one on F/mF among these, which are stored by the following instruction.

INSTRUCTION: Run through the finitely many automorphisms $Aut(F/mF)$, and see, whether $[f(''j)] \in F/jF$ is mapped onto $[f(j'')] \in F/\underline{j}F$ for $1 \leq j \leq s$. If this is the case, store the appearing automorphism.

Let STO denote the set of these stored automorphisms. If STO is empty, the decision of the proposition is answered negatively. Otherwise the following instruction gives the rest.

INSTRUCTION: Run through the stored automorphisms $\underline{C} : F/mF \to F/mF$ in STO, and decide (by the corollary above), whether $\underline{C}$ can be extended to an automorphism $C : F \to F$, which maps $'i$ onto i'.

If the answer is positive, the appearing automorphism has then obviously the wished properties. One has to scrutinize the present section from the beginning to see the following.

Observation

If the algorithm of the corollary above has given a positive decision, then the existing extension automorphism $A : TG \to TG$ is effectively calculable.

Theorem   (for comparing cohomology classes, X a finite complex)

Let $k, \underline{k} \in H^n(X; TG)$ be two cohomology classes given by representing cocycles $K, \underline{K} \in Z^n(X; TG) \subset \hom(C_n(X), TG) \cong \oplus TG$, i.e. by their values $K(\bar{i})$, $\underline{K}(\bar{i}) \in TG)$ on the non-degenerate n-simplces $\bar{i} \in X_n$, $1 \leq i \leq j$. It is effectively decidable, whether k can be transformed into $\underline{k}$ by a coefficient automorphism $\underline{C} : TG \to TG$.

Proof ($B^n(X; TG)$ the subgroup of coboundaries in $C^n(X; TG)$)
Because of the inclusion $H^n(X; TG) \to C^n(X; TG)/B^n(X; TG) = U(X; TG)$, we may consider instead of $k, \underline{k}$ the projections $u, \underline{u}$ of K resp. $\underline{K}$ in $U(X; TG)$. Let us diagonalize the boundary $d : C_n(X) \to C_{n-1}(X)$ into D. Applying the functor $hom(\text{-}, \text{identity})$ we get the dual diagonalizing of the coboundary d' into D', which is natural in TG

$$
\begin{array}{ccccc}
C^{n-1}(X; TG) & \xrightarrow[R']{\cong} & C^{n-1}(X; TG) & & \\
d' \downarrow & & \downarrow D' & & \\
C^n(X; TG) & \xrightarrow[S']{\cong} & C^n(X; TG) & \to & C^n(X; TG)/\text{image}(D') \\
& & & & = V(X; TG)
\end{array}
$$

the matrix S' maps $U(X; TG)$ isomorphicly onto $V(X; TG)$, and hence we may consider instead of $u, \underline{u}$ their images $v, \underline{v}$ in $V(X; TG)$.

This means, that we have two j-tuples $('1, '2, \ldots, 'j)$, $(1', 2', \ldots, j')$ of elements $'i, i' \in TG/\underline{i}TG$, $1 \leq i \leq j$, with $\underline{i} \geq 0$ integers (= the diagonal elements of D'), and we are looking for an automorphism $\underline{C} : TG \to TG$, which maps $'i$ onto i'.

We clearly may drop the indices i, where $\underline{i} = 1$, because then $TG/\underline{i}TG = 0$. For indices i, where $\underline{i} = 0$ it is $TG/\underline{i}TG = TG$. Hence we are exactly in the situation of the proposition above (after a trivial renumbering), and it is effectively decidable, whether an appropriate $C : TG \rightarrow TG$ exists. In the positive case C is effectively calculable by the observation above.

Application to k-Invariants

In the algorithm of section **IV** the calculated k-invariants appear as cohomology classes

$$u \in H^{i+1}(U; TG) , \quad v \in H^{i+1}(V; TG) , \quad G = \pi_i(F) ,$$

where U, V are terms of the preconstructible filtration of the stage $W = i - 1.P$, i.e. $U \subset V \subset W$, and the restrictions r and qr

$$H^{i+1}(W; TG) \xrightarrow{r} H^{i+1}(V; TG) \xrightarrow{q} H^{i+1}(U; TG)$$

are monomorphisms. The theorem above must be applied to the cohomology classes

$$u, \underline{u} = q(v) \in H^{i+1}(U; TG) .$$

Concluding Remarks

1. The algorithm of section **III**, which constructs Postnikov stages and calculates k-invariants cannot sensibly be considered as a further (= third) – the first one via iterated loop complexes, the second one via the functorial Postnikov tower – homotopy group algorithm: the homology of the Postnikov stages is needed and computed by an algorithm having the second one as a partial algorithm.

Exercise: Modify the algorithm of section **III** by using the cone of the composition $X \rightarrow n.X \rightarrow n-1.P$ instead the one of $n.X \rightarrow n-1.P$, and show, that the homology of the stages can be computed without [6] by using the fiberspace calculations of [9] directly. (Hence we have now another homotopy group algorithm working with finite subcomplexes of the canonical Postnikov-stages, which are minimal complexes. This algorithm can be considered as a constructive version of the classical Postnikov construction and comes near to the one of E.H. Brown [1], who had to prove extra a cumbersome, lengthy deformation lemma in order to have appropriate constructible, approximation complexes of the stages.)

2. It should be noted, that as with the Five-Lemma algorithm (see the 2. remark [5, p. 9]) the presentation can be lifted to a more abstract level, which would comprise the classical non-effective (i.e. for instance for infinite spaces) Postnikov-construction, too. In the new presentation the notion of algorithm does not occur; pure and applied viewpoints coincide. With help of the following key definition the idea might become clearer.

   Definition: An object $_iX$ (of some category of sequences of spaces, groups, or whatsoever, together with morphisms $_iX \rightarrow {_{i+1}}X$, $i \geq 1$) is called **stationary by**

function, if there exists a function $f : \mathbb{N} \to \mathbb{N} \times \mathbb{N}$, $f(i) = ('i, i')$, $i \leq 'i \leq i'$, such that for each $j \geq i'$ the canonical morphism

$$\text{image}(_iX \to _{i'}X) \to \text{image}(_iX \to _jX)$$

is an isomorphism.

The special cases 'f calculable, $_iX \to _{i+1}X$ constructible' and 'f the diagonal, $_iX$ all identical $= X$ arbitrary' yield then the applied resp. pure versions. Observe the relationship to the well known Mittag-Leffler condition, mostly defined for inverse directed systems of groups.

3. The following unsolved problem might give a flavour for the possibilities of application to very concrete examples.

   Open problem: Is the obstruction-index $o(f) \in H^m(M; \pi_{m-1}(F))$ for deforming a self-map $f : M \to M$ into a fixed point free map, effectively calculable? (see [10] for explanations, and observe, that $o(f)$ may be interpreted as the first possibly non-vanishing k-invariant of a certain fibration. A generalized Postnikov-decomposition with local coefficients must be used. Conjecture: yes, it is, if $\pi_1(M)$ is finite).

References

[1]   E.H. Brown, Finite computability of Postnikov-complexes, Ann. of Math.(2) 65 (1957), 1–20
[2]   K. Lamotke, Semisimpliziale algebraische Topologie, Grundl. der math. Wissenschaften 114, Springer, Berlin 1968
[3]   J.P. May, Simplicial objects in Algebraic Topology, Van Nostrand Company, New York 1969
[4]   M. Newman, Integral matrices, Academic Press, New York and London 1972
[5]   R. Schön, A Five Lemma for Calculations in Homological Algebra, Memoirs of the AMS, this issue
[6]   R. Schön, An Algorithm for Calculating Homotopy Groups, Memoirs of the AMS, this issue
[7]   K. Shoda, Über die Automorphismen einer endlichen abelschen Gruppe, Math. Ann. 100 (1928), 674–686
[8]   J.R. Hubbuck, On homotopy-commutative H-spaces, Topology 8, 1969, 119–126
[9]   R. Schön, Fibrations with Calculable Homology, Memoirs of the AMS, this issue
[10] E. Fadell, S. Husseini, Fixed point theory for non simply connected manifolds, Topology 20, 1981, 53–92
[11] T. tom Dieck, K.H. Kamps, D. Puppe, Homotopietheorie, Lecture Notes in Mathematics No. 157, Springer, Berlin (1970)
[12] G.W. Whitehead, Elements of Homotopy Theory, Springer, New York, Berlin, Heidelberg 1978

Friedensstr. 2
6900 Heidelberg
Fed. Rep. of Germany

MEMOIRS of the American Mathematical Society

SUBMISSION. This journal is designed particularly for long research papers (and groups of cognate papers) in pure and applied mathematics. The papers, in general, are longer than those in the TRANSACTIONS of the American Mathematical Society, with which it shares an editorial committee. Mathematical papers intended for publication in the Memoirs should be addressed to one of the editors:

Ordinary differential equations, partial differential equations and applied mathematics to ROGER D. NUSSBAUM, Department of Mathematics, Rutgers University, New Brunswick, NJ 08903

Harmonic analysis, representation theory and Lie theory to AVNER D. ASH, Department of Mathematics, The Ohio State University, 231 West 18th Avenue, Columbus, OH 43210

Abstract analysis to MASAMICHI TAKESAKI, Department of Mathematics, University of California, Los Angeles, CA 90024

Real and harmonic analysis to DAVID JERISON, Department of Mathematics, M.I.T., Rm 2–180, Cambridge, MA 02139

Algebra and algebraic geometry to JUDITH D. SALLY, Department of Mathematics, Northwestern University, Evanston, IL 60208

Geometric topology and general topology to JAMES W. CANNON, Department of Mathematics, Brigham Young University, Provo, UT 84602

Algebraic topology and differential topology to RALPH COHEN, Department of Mathematics, Stanford University, Stanford, CA 94305

Global analysis and differential geometry to JERRY L. KAZDAN, Department of Mathematics, University of Pennsylvania, E1, Philadelphia, PA 19104-6395

Probability and statistics to RICHARD DURRETT, Department of Mathematics, Cornell University, Ithaca, NY 14853-7901

Combinatorics and number theory to CARL POMERANCE, Department of Mathematics, University of Georgia, Athens, GA 30602

Logic, set theory, general topology and universal algebra to JAMES E. BAUMGARTNER, Department of Mathematics, Dartmouth College, Hanover, NH 03755

Algebraic number theory, analytic number theory and modular forms to AUDREY TERRAS, Department of Mathematics, University of California at San Diego, La Jolla, CA 92093

Complex analysis and nonlinear partial differential equations to SUN-YUNG A. CHANG, Department of Mathematics, University of California at Los Angeles, Los Angeles, CA 90024

All other communications to the editors should be addressed to the Managing Editor, DAVID J. SALTMAN, Department of Mathematics, University of Texas at Austin, Austin, TX 78713.

General instructions to authors for

PREPARING REPRODUCTION COPY FOR MEMOIRS

> **For more detailed instructions send for AMS booklet, "A Guide for Authors of Memoirs."**
> **Write to Editorial Offices, American Mathematical Society, P.O. Box 6248,**
> **Providence, R.I. 02940.**

MEMOIRS are printed by photo-offset from camera copy fully prepared by the author. This means that the finished book will look exactly like the copy submitted. Thus the author will want to use a good quality typewriter with a new, medium-inked black ribbon, and submit clean copy on the appropriate model paper.

Model Paper, provided at no cost by the AMS, is paper marked with blue lines that confine the copy to the appropriate size.

Special Characters may be filled in carefully freehand, using dense black ink, or **INSTANT** ("rub-on") **LETTERING** may be used. These may be available at a local art supply store.

Diagrams may be drawn in black ink either directly on the model sheet, or on a separate sheet and pasted with rubber cement into spaces left for them in the text. Ballpoint pen is not acceptable.

Page Headings (Running Heads) should be centered, in CAPITAL LETTERS (preferably), at the top of the page — just above the blue line and touching it.

LEFT-hand, EVEN-numbered pages should be headed with the AUTHOR'S NAME;

RIGHT-hand, ODD-numbered pages should be headed with the TITLE of the paper (in shortened form if necessary).

Exceptions: PAGE 1 and any other page that carries a display title require NO RUNNING HEADS.

Page Numbers should be at the top of the page, on the same line with the running heads.

LEFT-hand, EVEN numbers — flush with left margin;

RIGHT-hand, ODD numbers — flush with right margin.

Exceptions: PAGE 1 and any other page that carries a display title should have page number, centered below the text, on blue line provided.

FRONT MATTER PAGES should be numbered with Roman numerals (lower case), positioned below text in same manner as described above.

MEMOIRS FORMAT

> It is suggested that the material be arranged in pages as indicated below.
> Note: <u>Starred items (*)</u> are requirements of publication.

Front Matter (first pages in book, preceding main body of text).

Page i — *Title, *Author's name.

Page iii — Table of contents.

Page iv — *Abstract (at least 1 sentence and at most 300 words).

Key words and phrases, if desired. (A list which covers the content of the paper adequately enough to be useful for an information retrieval system.)

*1991 Mathematics Subject Classification. This classification represents the primary and secondary subjects of the paper, and the scheme can be found in Annual Subject Indexes of MATHEMATICAL REVIEWS beginnning in 1990.

Page 1 — Preface, introduction, or any other matter not belonging in body of text.

Footnotes: *Received by the editor date.
Support information — grants, credits, etc.

First Page Following Introduction – Chapter Title (dropped 1 inch from top line, and centered). Beginning of Text.

Last Page (at bottom) – Author's affiliation.